Near Vertical Incidence Skywave Communication

Theory, Techniques and Validation

by

LTC David M. Fiedler, (NJ ARNG) (Ret)

and

LTC Edward J. Farmer, P.E. (CA SMR), AA6ZM

Near Vertical Incidence Skywave Communication
Theory, Techniques and Validation

by David M. Fiedler and Edward J. Farmer

Copyright October, 1996

Editor Biographical Data

LTC David M. Fiedler

LTC Fiedler was commissioned in the U.S. Army Signal Corps upon graduation from the Pennsylvania Military College (now Widener University) in 1968. He is a graduate of the Signal Officers Basic Course, the Radio and Microwave Systems Engineering Course, the Signal Officers Advanced Course, the Military Intelligence Officers Course, and the U.S. Army Command and General Staff College. He has served in Regular Army, Army Reserve, and Army National Guard signal, infantry, and armor units in CONUS and in Vietnam.

LTC Fiedler holds degrees in physics and engineering and an advanced degree in industrial management. His last military assignment was as the Chief of the Communications – Electronics Division of the New Jersey National Guard State Area Command where he also served on the New Jersey Emergency Communications Commission. He also holds a General Class amateur radio license call sign WB2CDG.

Concurrent with his military career, LTC Fiedler was employed for 37 years as a senior Department of the Army civilian engineer and technical manager. He has been a systems engineer/project leader for many Army communications and electronic warfare projects and programs including, the Joint Tactical Fusion Program (Asst Program Manager), All Source Analysis System (ASAS), Mobile Subscriber Equipment (MSE), Theater Nuclear Forces (TNF) communications, Single Channel Ground and Airborne Radio Communications Systems (SINCGARS), Transformation High Frequency Radio (THFR), to name but a few.

After retirement in 2005 he has been a consultant for various government, military, professional, academic, and business organizations.

LTC (CA) Edward J. Farmer, P.E.

Ed Farmer graduated from the California State University at Chico in 1971 where he earned degrees in electrical engineering and physics. He is a Registered Professional Engineer and president of EFA Technologies, Inc.

He completed signal school at Ft. Gordon during the Viet-Nam war and worked in Military Intelligence. He was assigned to the Directorate of Plans, Operations, and Security of the California National Guard where he served as Assistant State Signal Officer. He completed U.S. Marine Corps Command and Staff College on 27 October 1999.

He earned his original Amateur Radio License in 1959 (WL7DNZ) and presently holds Amateur Extra Class License AA6ZM.

He has been awarded five patents, has authored four books, and published over forty papers and articles.

THIRD EDITION

OCTOBER 2017

LICENSED PUBLISHER
JOKALYM PRESS
CULPEPER VA
k3mt@arrl.net

CONTENTS

INTRODUCTION

INTRODUCTION

Radio waves are part of the electromagnetic spectrum which extends from near direct current through light. For convenience, various ranges of frequencies with somewhat similar propagation characteristics are referred to in terms of their place in the electromagnetic spectrum.

Medium frequencies are those in the range of 300 kHz through 3 MHz. Services that operate in this range include (among others) AM radio broadcast stations, radio location beacons (LORAN), some military communications, and Amateur Radio. The high frequency range extends from 3 MHz through 30 MHz and includes many services such as ship-to-shore radio, international broadcast services, military communications, commercial services, and Amateur Radio.

Medium and high frequency radio has been around for a long time. Marconi's experiments, which introduced the world to long range radio communication, were conducted in the high frequency range. Radio Amateurs regularly demonstrate the ability to work hundreds of countries by using their frequencies between 1.8 and 30 MHz. Ship radiotelephone service has provided many mariners and their passengers with telephone service across the oceans.

Above the high frequency range there are very high frequency (VHF) which encompasses 30 to 300 MHz, ultra high frequency (UHF) which includes 300 to about 3000 MHz, super high frequency (3 to 30 GHz), extremely high frequency (30 GHz to 300 GHz) and still higher ranges which range into the terahertz region.

Most commercial and military interest is in these higher frequencies. There are several reasons for this but perhaps the most important is bandwidth. The mathematician C.E. Shannon established in 1948 that our ability to move information over radio is related to the bandwidth available. The entire medium and high frequency range includes a bandwidth of only 29.7 MHz. The VHF range alone has a bandwidth of 270 MHz — *greater by nearly ten times*. The UHF range has a bandwidth of about 2,700 MHz, over ninety times the bandwidth

available at medium and high frequency. Do you see a trend emerging here? Rapid transmission of large volumes of data requires bandwidth which is increasingly available with increasing frequency.

Many modern applications of communication technology require huge amounts of bandwidth. For example the entire high frequency band would support only four television channels! Similar loading is created by high speed data networks. Clearly, high speed *and* high volume favors use of higher frequencies.

Obviously, medium and high frequency radio encompass only a tiny portion of the radio spectrum. Yet, as long as the limitations are kept in mind, it can do amazing things which can not be duplicated at any other frequencies. One of those, Near Vertical Incidence Skywave (NVIS) communication, is of significant importance in emergency and tactical military communications. This mode has traditionally been misunderstood and poorly appreciated. Hopefully the next hundred or so pages will help that situation.

NVIS propagation occurs readily at the high end of the medium frequency range and in the lower half of the high frequency range. Radio Amateurs find their 160-meter band (1.8 to 2 MHz), 80-meter band (3.5 to 4 MHz) and 40-meter band (7.0 to 7.3 MHz) to be useful. Military applications typically range from 2 MHz to about 12 MHz.

At frequencies between 1.8 and 30 MHz there are three propagation modes: line-of-sight, ground wave, and sky wave.

Line of sight propagation occurs between stations that have visibility with each other. While high frequency radio waves refract somewhat as they pass through the atmosphere paths of this type are essentially limited to stations that have visibility with each other. This is the most common propagation mode at VHF, UHF, and above.

Ground wave propagation results from radio waves following the surface of the earth. In this mode, they are guided waves and, depending on the earth under them, can travel dozens of miles to reach their receiver. This is the usual mode by which we hear AM broad-

Bandwidth in Various Frequency Ranges		
Band	**Frequency Range**	**Bandwidth (MHz)**
medium frequency (MF)	0.3 to 3 MHz	2.7
high frequency (HF)	3 to 30 MHz	27
very high frequency (VHF)	30 to 300 MHz	270
ultra high frequency (UHF)	300 to 3000 MHz	2,700
super high frequency (SHF)	3 to 30 GHz	27,000
extremely high frequency (EHF)	30 to 300 GHz	270,000

cast stations. While this mode exists at VHF and above it is most common at medium frequencies (300 kHz to 3 MHz) and high frequencies (3 MHz to 30 MHz).

Skywave propagation involves reflecting signals off the ionosphere. It is in this mode that medium and high frequency radio exhibits its unique and special qualities. This reflection technique allows us to leap tall mountains with a single bound. It enables us to talk with stations on the other side of the earth. And, if used correctly, enables us to provide continuous and dependable coverage of areas of operation that span several hundred miles. It is this capability, and the need for it in tactical operations of regional agencies as well as military corps and smaller units, that makes it important to understand NVIS communications techniques.

Most of the good NVIS literature is unclassified and in the public domain but it is not easy to access. Probably 90 percent of it has been published in *Army Communicator* magazine, an official publication of the U.S. Army Signal Corps. The editors felt the public would benefit by access to the better of these publications and set out to develop this book. Hopefully, this will place the best information about NVIS within reach of the commercial, civilian, and Amateur Radio communities.

Readers are also advised to explore the works of Dr. Alan Christinsin. His two-volume set, *Tactical HF Radio Command and Control — an Anthology,* is excellent. It is available from the author at ASC & Associates, Ltd., 1201 Dawn Dr., Belleville, IL 62220.

In Part I, the basis for NVIS communication is explained. Part II covers how to do it, and Part III presents confirmation that it works. There is some unavoidable overlap among these sections but hopefully it isn't unduly distracting. Together, these three sections build a compelling case for the use of high frequency radio for certain missions.

The editors sincerely hope you will find this book useful.

PART ONE

HOW AND WHY NVIS WORKS

Near Vertical Incidence Skywave communication occurs when signals of appropriate frequency are properly directed toward, and reflected from, the ionosphere. The requirements for success include proper frequency selection and suitable antenna design. This collection of papers provides the needed technical background and some practical information regarding how to deal with both issues.

1. Beyond Line-of-Sight Propagation and Antennas by LTC David M. Fiedler, **Army Communicator** magazine, Fall 1983.

2. Skip the Skip Zone by LTC David M. Fiedler, **Army Communicator** magazine, Spring 1986.

3. NVIS Propagation at Low Solar Flux Indices by Maj Edward J. Farmer, **Army Communicator** magazine, Spring 1994.

4. NVIS Antenna Fundamentals by MAJ Edward J. Farmer, **Army Communicator** magazine, Fall 1994.

Beyond line-of-sight propagation modes and antennas

by David M. Fiedler and George H. Hagn

The US Army has utilized high frequency (HF) radio (2-30 MHz) for both strategic and tactical communications purposes for approximately the last 60 years in order to achieve ranges beyond line-of-sight (BLOS). In particular, HF was used in WWII, the Korean War, and the war in Southeast Asia to provide tactical communications over BLOS ranges in difficult terrains (mountains and jungles, for example). Initially, the mid-1970s Integrated Tactical Communications Systems (INTACS) study showed existing tactical HF links being replaced by satellite links. A more recent recognition of the cost, vulnerabilities, and availability of satellites, and HF technology advances have led to a renewed recognition of the military utility of modern HF systems for satisfying these requirements. Additionally, in today's tactical units, HF radio is a major means of communication which must be used in any "come-as-you-are war." Therefore, it is timely to reconsider the propagation modes and antennas recommended for tactical use. Over the years, the theory of HF communications and its military applications have been described in various US Army technical and field manuals which today have culminated in publications such as FM 24-18 (Field Radio Techniques), FM 11-65 (High Frequency Radio Communications) and TM 11-666 (Antennas and Radio Propagation). All of these publications place primary emphasis on what in the past have been the most useful modes of HF radio propagation for military purposes. These modes are useful for short range (0-50 miles) tactical communication (see figure 1) and for long range communication (see figure 2) and include modes such as:

Direct Wave Mode: Defined as the component of a wave front that travels directly via line-of-sight (LOS) from the transmitting antenna to the receiving antenna (see figure 1). This mode is very useful for air-to-ground communications since terrain features which absorb RF energy normally are not in the propagation path. Over the ground, however, the range for this mode is limited by terrain absorption and by path blockage caused by terrain features.

Ground-Reflected Wave Mode: Defined as the component of the radiated wave that reaches the receiver after being reflected from the ground (see figure 1). Since the ground-reflected wave travels a longer distance, it arrives at the receive antenna later than the direct wave, and it can cancel or enhance the direct-energy waves of depending upon the geometry, frequency, and the reflection coefficient of the ground. The ground-reflected wave can be used for communications under some circumstances if the reflected wave is somehow less attenuated than the direct wave at the receiver — although most of the time, it is undesirable since it tends to weaken the direct wave for most tactical geometries at HF.

Space Wave Mode: Defined as the combination (the vector sum) of the direct and ground-reflected waves. Militarily, the phenomena associated with the space wave led to much maneuvering on the battlefield to assure that the communications equipment was sited on the highest ground in order to reduce the effect of terrain on range and effectiveness. The space wave provides the best propagation mode for short-range LOS tactical HF communications once the location problems are overcome.

Surface Wave Mode: Defined as that component of the groundwave (see e below) that travels along the earth's surface (see figure 1) and is primarily affected by the conductivity and/or dielectric constant of the earth. When the transmitting and receiving antennas are located close to the earth (as they are in most tactical communications applications), the direct and ground-reflected waves tend to cancel each other. In this case, the resulting composite signal is principally that of the surface wave. The surface wave diminishes in strength with height above the ground, and usually it is not very useful above about one wavelength over the ground. The energy of the surface wave is absorbed by the earth at twice the rate of the direct wave mode (in dB) as it travels over the ground. A distance is reached where it can no longer be used for communication. This usable distance over the earth's surface can be increased by polarizing the wave in a vertical orientation since the earth produces much less of an attenuation effect on vertical than on horizontal polarization for antennas located near the earth.

Groundwave Mode: Defined as the vector sum of the space wave and the surface wave. This formulation of the theory of groundwave propagation by A. Sommerfeld in the early 1900s was made practical for engineering use by K. A. Norton in the late 1930s and early 1940s. As stated in FM 24-18, under ideal conditions, useful groundwave energy can extend to ranges up to 50

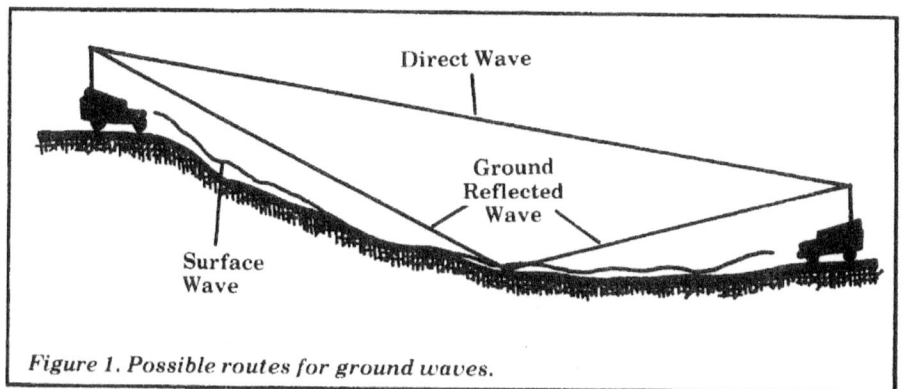

Figure 1. Possible routes for ground waves.

miles. This is well BLOS; however, there are much more common conditions which limit useful groundwave propagation to as few as 2 miles, for example, manpack radio operations in a wet, heavy jungle. This wide variation in range is due to the varying condition of the earth (ground conductivity, vegetation, terrain irregularity, and so on), the atmosphere, and radio noise; and thus it cannot be controlled by the communicator. This can be a disaster when trying to establish effective communications for command and control (C^3) in units deployed over a wide area when the communications planner does not properly consider the impact of these effects on the performance of his communications systems.

Classical Skywave Mode: Skywave is defined as those types of radio transmission that make use of ionospheric reflection. Skywave modes can provide communication over longer distances than can be achieved via groundwave. The reflection of HF energy from the ionosphere back to earth is dependent upon such things as the number of free electrons per unit volume in the ionosphere, the height of the ionized layer, operating frequency, incidence angle, etc. These variables, however, are not the main subject of this article. A good detailed description of this mode is given in TM 11-666 and in reference six at the end of this essay.

Figure 2 (see also references 1-3 at the end of this paper) depicts the skip-zone problem that faces tactical communicators. Under ideal conditions, the groundwave becomes unusable at about 50 miles. Under actual field conditions, this range can be much less, again sometimes as few as 2 miles. Successful skywave communication for any length of path and system depends upon the selection of a frequency which is low enough to be reflected by the ionosphere. The selection of the proper antennas is also important. Whips are commonly used, and they have low gain for skywaves on such short paths. Therefore, frequency selection and/or antenna choice can leave a skip zone of at least 50 (and more probably 70) miles where HF communication will not function. Translating this into terms of military deployments, this means that units such as long-range patrols, armored cavalry deployed as advanced or covering forces, air-defense early-warning teams, and many division-CORPS, division-BDE, division-DISCOM, division-DIVARTY stations using whips are in the skip zone. Thus

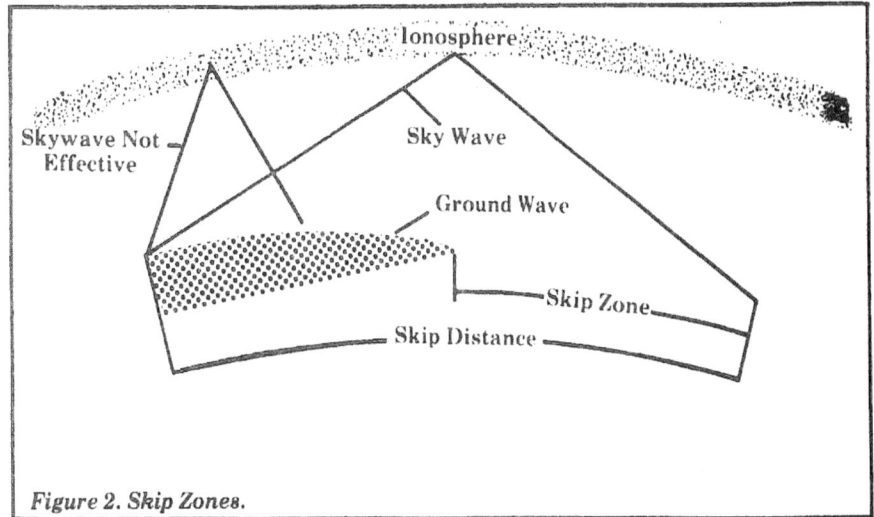

Figure 2. Skip Zones.

they are unreachable by HF radio under skip zone conditions even though HF is a primary means of communication planned for use by these units.

A closer examination of figure 2 shows a wave striking the ionosphere at a high angle and being reflected into an area covered by a strong groundwave signal. This wave is labeled "skywave not effective" in all references because the groundwave signal strength is much stronger than the skywave signal. Unfortunately, the figure is misleading in several ways. Energy radiated in a near-vertical-incidence direction is not reflected down to a

pinpoint on the earth's surface. If it is radiated on too high a frequency, the energy penetrates the ionosphere and continues on out into space. Energy radiated on a low enough frequency is reflected back to earth at all angles (including the zenith) resulting in the energy striking the earth in an omnidirectional pattern without dead spots (without a skip zone) if an efficient short-path antenna such as a doublet is used. Such a mode is called a near-vertical-incidence skywave (NVIS) mode. This mode is shown in figure 2, but the concept is illustrated in figure 3. This effect is similar to taking a hose

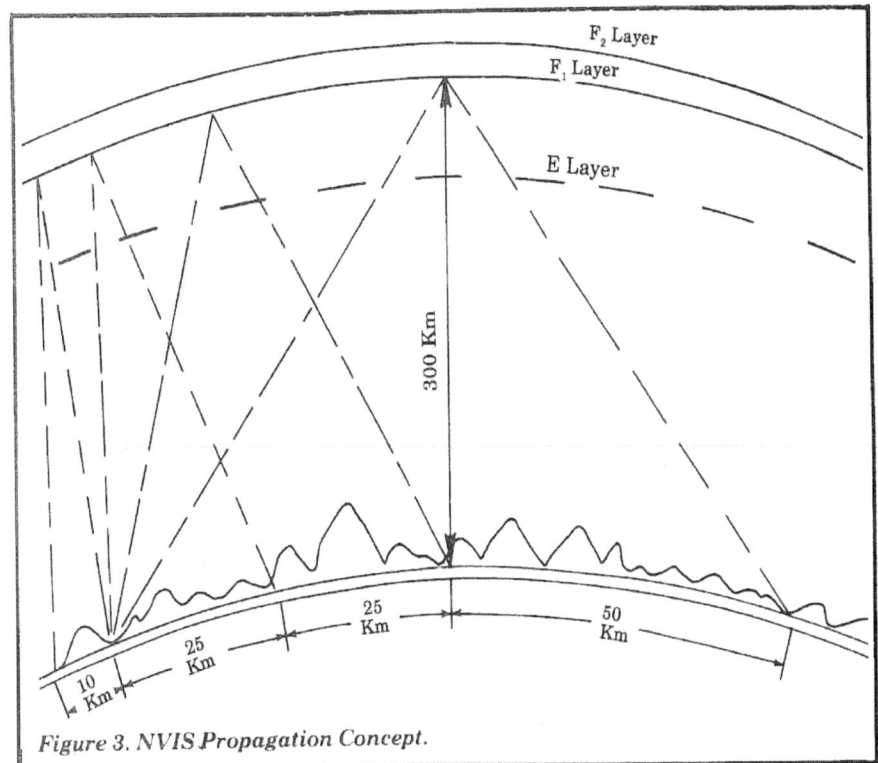

Figure 3. NVIS Propagation Concept.

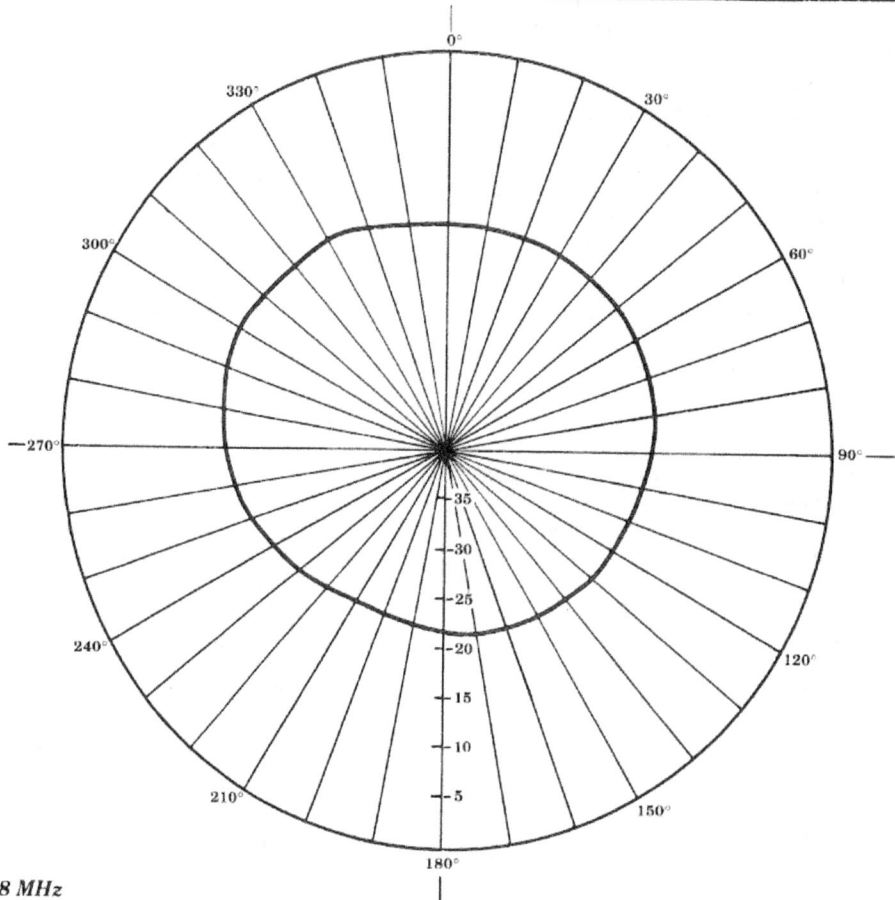

10 May 74
11:30 Hours
-18.5 dB relative to
half-wave horizontal
dipole

(Ricciardi and Brune, 1979)

Figure 4. NVIS Antenna Pattern 4.328 MHz

with a fog nozzle and pointing it straight up. The water falling back to earth covers a circular pattern continuously out to a given distance. A typical NVIS received-signal pattern is shown in figure 4, and the path is shown in figure 5. The main difference between this short-range NVIS mode and the standard long-range skywave HF mode is the lower frequency required to avoid penetrating the ionosphere at the near-vertical angle of incidence of the signal upon the ionosphere. In order to attain an NVIS effect, the energy must be radiated strongly enough at angles greater than about 75 or 80 degrees from the horizontal on a frequency that the ionosphere will reflect at that location and time. The ionospheric layers will reflect this energy in an umbrella-type pattern with no skip zone. Any groundwave present with the NVIS skywave signal will result in undesirable wave interference effects (fading) if the amplitudes are comparable. However, proper antenna selection will reduce groundwave radiated energy to a

minimum, and this will reduce the fading problems. Ranges for the NVIS mode are shown in figure 5 for a typical ionosphere height (300 km) and takeoff angles. Since NVIS paths are purely skywave, the path losses are nearly constant at about 110dB +10dB. This is significant for the tactical communicator since all the energy arriving at his receive antenna is coming from above at about the same strength over all of the communications ranges of interest. This means the effects of terrain and vegetation (when operating from defiladed positions such as valleys) are greatly reduced, and the receive signal strength will not vary greatly with relatively small changes in location. This is especially important for helicopters flying nap-of-the-earth beyond VHF radio range.

The need for short-range HF communications without skip zones is obvious. Therefore, our next problem is how to generate the required radiation characteristics. Fortunately, this is not difficult since 1/2-wave dipole antennas

located from 1/4 to 1/10 wavelength above the ground will cause the radiated energy to be directed vertically. Table 1 shows the relative gain toward the zenith of the most common types of HF field-expedient antennas. This table shows that the 1/2-wave Shirley Folded Dipole (see figure 6) has the most gain toward the zenith (with the other dipoles being almost as good). The Shirley dipole is a good NVIS base station antenna, but it is limited to a band of frequencies within about 10 percent of the design frequency. The fan dipole (see figure 6 and table 1) performs almost as well, and it provides more frequency flexibility (day, night and transition period frequencies). For tactical communications, these dipoles can be easily deployed in a field-expedient manner because they can be located close to the ground. For mobile (or shoot and scoot) operations, vehicle-mounted antennas are required. The answer to this problem is the standard 16 1/2-foot whip bent down into a horizontal position. In this configuration,

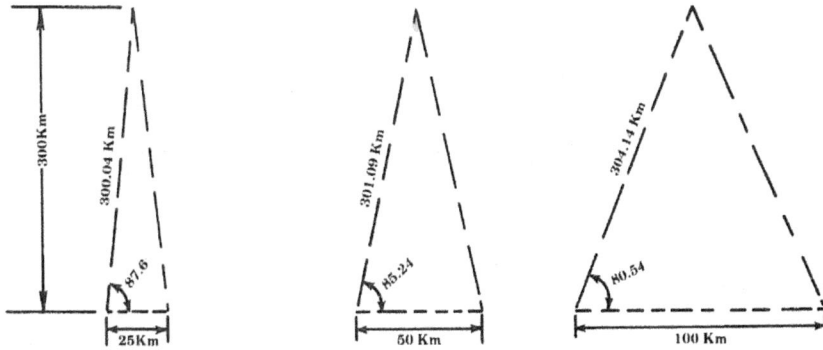

Ground Range Km	Radio Path Length Km	Range Variable 20 log d	V Loss dB
25	600.5	55.57	0
50	602.08	55.59	.02
100	608.28	55.68	0.11

Night or Day absorption, turn-around, normal atmospherics.
Average Path Loss (3 to 5 MHz) + 110 dB + 10 dB.

Figure 5. Path Length and Incident Angle (NVIS Mode.

the whip is essentially an asymmetrical dipole (with the vehicle body forming one side) located close to the earth, with a significant amount of energy being directed upward to be reflected back by the ionosphere in an umbrella pattern. For use while operating on the move, of course, the whip antenna must be tied across or parallel to the vehicle or shelter. This configuration is more like an asymmetrical open-wire transmission line, and it also will direct some energy upward — although with less efficiency. There are still no skip zones with proper frequency selection, but received signal levels are weaker than with the whip tied back. Special NVIS antennas designed primarily for

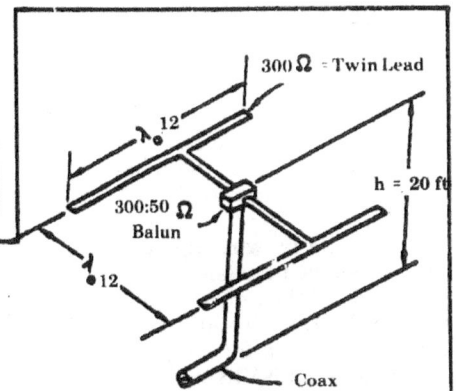

(A) Shirley Dipole Array

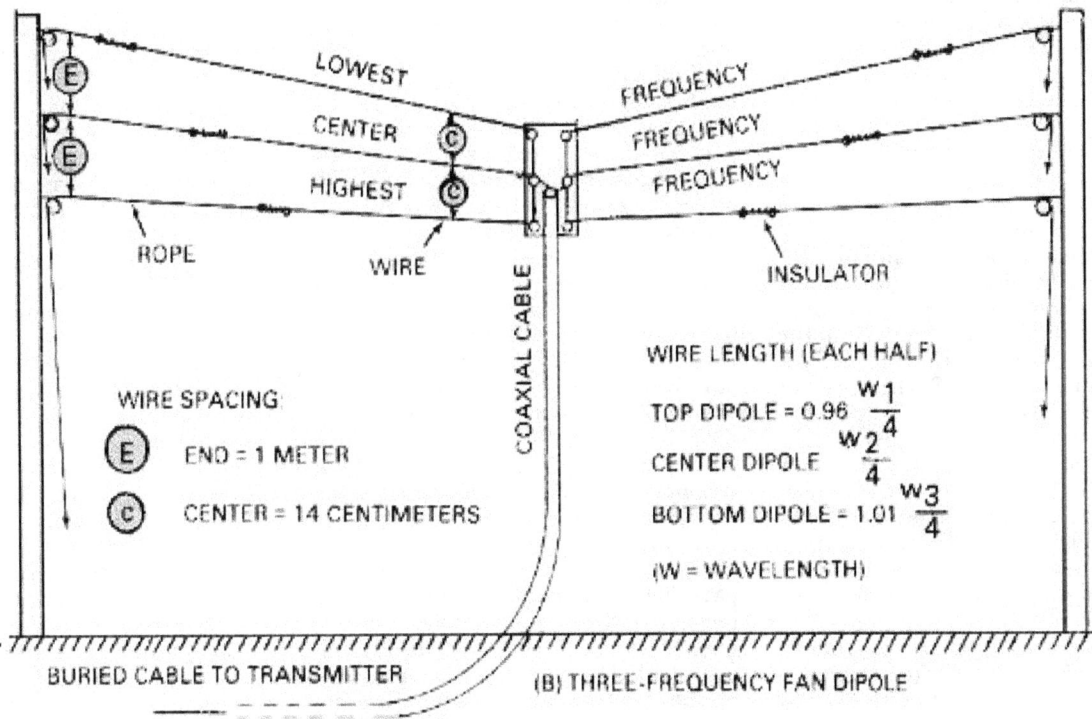

WIRE SPACING:

(E) END = 1 METER

(C) CENTER = 14 CENTIMETERS

WIRE LENGTH (EACH HALF)

TOP DIPOLE = 0.96 $\dfrac{w_1}{4}$

CENTER DIPOLE $\dfrac{w_2}{4}$

BOTTOM DIPOLE = 1.01 $\dfrac{w_3}{4}$

(W = WAVELENGTH)

(B) THREE-FREQUENCY FAN DIPOLE

Figure 6. Shirley and Fan Dipole NVIS Base Station Antennas.

Antennas	Clearing	75-ft forest clearing	50-ft forest clearing		
h/2 unbalanced single-wire dipole	+1.0	-2.8	0.0	-1.2, -1.7	0.0
h/2 balanced single-wire dipole	+0.5	-3.7	no data	no data	no data
h/2 folded dipole (30-0:5-0 U balum)	+0.2	-1.0	no data	no data	no data
h/4 short (loaded to h/2) dipole	-3.0	-5.2	no data	no data	no data
h/2 sleeve dipole (on ground)	-32.1	-28.3	no data	no data	no data
3-freq. fan dipole @ 15 ft	-0.4	-5.1	no data	no data	no data
3-freq. fan dipole @ 12	-2.4	-5.6	no data	no data	no data
3-freq. fan dipole @ 9	-4.0	-8.1	no data	no data	no data
shirley folded dipole	+3.0	-0/3	no data	no data	no data
3 h/4 inverted L (1:h = 2:1)	-0.0	-2.8	no data	no data	no data
3 h/4 inverted L (1:h = 3:1)	-0.8	-3.3	no data	no data	no data
3 h/4 inverted L (1:h = 4:1)	-1.0	-5.8	no data	no data	no data
3 h/4 inverted L (1:h = 5:1)	-2.0	-6.3	-10.2	-10.7, -12.5	-9.0
30° slant wire (h/r elevated)	-10.1	-14.8	-11.8	-13.5, -14.2	-14.0
60° slant wire (h/r elevated)	-11.8	-14.8	no data	no data	no data
10-ft square (vertical plane) loop @ 6 ft	-24.1	-25.3	no data	no data	no data
16.5-ft whip	-41.5	-44.0	-31.7	-25.0, -25.2	no data

Summary of relative gain toward the Zenith for Field-expedient HF antennas

Table 1.

helicopters are also useful for this application, and they can be modified for shelter and ground vehicular operation.

Traditionally, wire dipole antennas have always been sited so that the broadside of the antenna was pointed toward the receive station. This is the correct approach for long-haul paths. When using the NVIS mode, this antenna orientation is unnecessary. For NVIS operation, the antenna orientation does not matter since all the energy is directed upward and returns to earth in what is essentially an omnidirectional pattern. In operational terms this means that the dipole should be erected at any orientation that is convenient at the particular radio site without regard to the bearing of other stations. This holds true except when operating in the region of the "magnetic dip equator." When operating within 500 km of the dip equator, the dipole antennas should be oriented in a magnetically north-south direction for greater received signal levels for all NVIS path bearings. US Army Special Forces made use of this dipole north-south orientation in their HF single sideband (SSB) net in the Mekong delta during the Vietnam War with excellent results. Traditional antenna orientation (broadside to the path direction) must be retained when operating on longer skywave paths near the dip equator and elsewhere.

While use of the NVIS technique does provide BLOS "skip-zone-free" communication, there are some drawbacks to its use that must be understood in order to minimize them. These include:

Interference between ground-wave and skywave: Both an NVIS and groundwave signal are present the groundwave can cause destructive interference. Proper antenna selection will suppress groundwave radiation and minimize this effect while maximizing the amount of energy going into the NVIS mode.

High-takeoff angles: In order to produce radiation which is nearly vertical (i.e., NVIS), antennas must be selected and located carefully in order to minimize the groundwave radiation and maximize the energy radiated towards the zenith. This can be accomplished by using specially designed antennas or by locating standard dipole (doublet) antennas 1/4 to 1/10 wavelength from the ground in order to direct the energy toward the zenith.

Frequency selection: In skywave propagation, there is a critical frequency (f_0) above which radiated energy generally will not be reflected by the ionosphere but will pass through it. This frequency is related approximately (by a constant k slightly greater than unity which depends primarily on path length) to the angle of incidence (o) and the classical maximum usable frequency (MUF) by the equation: MUF $= k\, f_0 \sec(o)$.

This means that the useful frequency range varies in accordance with the path length: the shorter the path the lower the MUF and smaller the frequency range. The lowest useful frequency (LUF) is determined primarily by the effective radiated power and the noise and interference at the receiver. Practically speaking, this limits the

NVIS mode of operation to the 2-4 MHz range at night and between 4-8 MHz during the day. These nominal limits will vary with the 11-year sunspot cycle and they will be smaller during sunspot minimums (1985-86 for example). Figure 7 is an example of the percent of time an operating frequency would have exceeded the maximum usable frequency (MUF) during a solar minimum. This restricting of the frequency range is due to the physics of the situation, and it cannot be overcome by engineering. Therefore, problems can be expected when using on the NVIS mode in the low end of the HF spectrum.

These problems include:

The range of frequencies between the MUF and the LUF is limited, and frequency assignment may be a problem — especially during the minimum part of the 11-year solar cycle when many users are crammed into the smaller available HF spectrum.

The lower portion of the band which supports NVIS is somewhat congested with aviation, marine, broadcast, and amateur users which limits frequencies available — even during the solar maximum.

Atmospheric noise is higher in this portion of the HF spectrum in the afternoon and at night.

Man-made noise tends to be higher in this portion of the HF spectrum.

All of these drawbacks of NVIS transmission, except the limited frequency range, can be overcome with relative ease. Once this is done, the many advantages to the tactical communicator are clear. They include:

Skip-zone-free omnidirectional communications.

Terrain-independent path loss resulting in a more constant received signal level over the entire tactical operational range instead of widely varying path loss with distance, and the corresponding uncertainty in operational range.

Capability of operating from defiladed positions eliminating the restriction on the tactical commander to control the high ground for HF communications purposes.

Non-critical antenna orientation of doublets and other linear antennas such as inverted L's.

Several electronic warfare advantages. First, lower probability

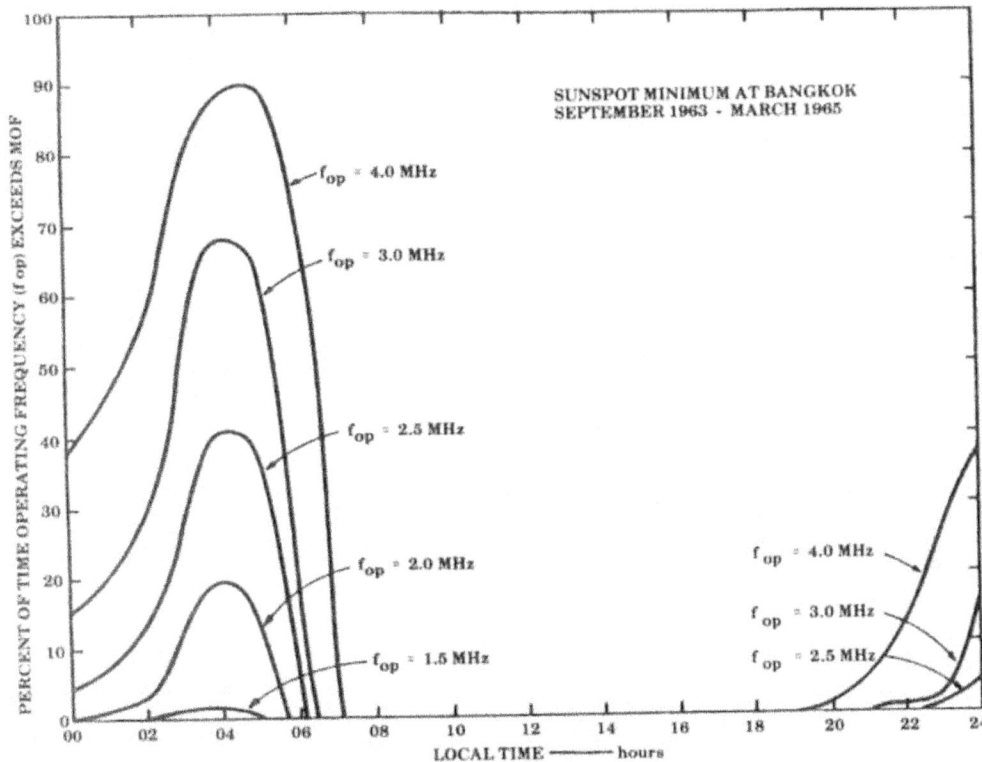

Figure 7. Percent of time the Operating Frequency Exceeds MOF of an NVIS
path of length less than 50 KM

of geolocation: NVIS is received from above at very steep angles. This makes direction finding (DF) from nearby (but beyond groundwave range) locations more difficult. Next, harder to jam with groundwave jammers (the most common tactical HF jamming mode) are subject to path loss. Since all NVIS radiated energy arrives from above (skywave), terrain features can be used to attenuate a groundwave jammer without degrading the desired communication path. When operating against a station using NVIS propagation, the jamming signal will be attenuated by terrain while the skywave NVIS path loss will be constant. This forces the groundwave jammer to move very close to the target or put out more power, and either tactic makes jamming more difficult. Finally, it requires only low-power operation: the NVIS mode can be used successfully (due to the constant path loss of tolerable size) with very low-power sets provided that proper frequency and antenna selections are made. This will result in much lower probabilities of

intercept and detection. Figure 8 shows results obtained in Thailand jungles and mountains with the 15-W AN/PRC-74 operating with selected antennas on one SSB voice frequency (3.6 MHz) over a 24-hour period during the 1963 sunspot minimum. Clearly the 1/2 wave dipole provided the best results, and it was operating NVIS for ranges beyond about 5 miles. The whip was operating groundwave out to about 20 miles and NVIS beyond, and the slant wire was intermediate between the dipole and whip. The performance with the dipole would have been even better if a frequency change (QSY) had been permitted near dawn (see figures 8 and 9), but QSY was not permitted in the test.

With this date in mind, tactical communicators should add another dimension to their thinking and planning. NVIS techniques must be considered under the following conditions:

When the area of operations is not conducive to groundwave HF communications (e.g., mountains).

When tactical deployments that place stations in anticipated skip

zones when using traditional antennas (whips), frequency selection methods and operating procedures.

When operating in heavy, wet jungles (or other areas of high signal attenuation).

When prominent terrain features are not under friendly control.

When operating from defiladed positions.

When operating against enemy groundwave jammers and direction finders.

When flying close to the ground in helicopters or in light aircraft.

Along with the addition of the NVIS technique to our tactical HF communication thinking, it is also necessary to amend our training and doctrine to reflect more completely all HF modes available to the communicator. The Air Force has incorporated some of this information into their literature, but we have been unable to find any Army TM, FM, or POI which properly describes the advantage of the short-path skywave techniques discussed in this paper. In all cases, these techniques are either ignored or down played. In the past, this situation was unfortunate, but it was tolerable since the ground-

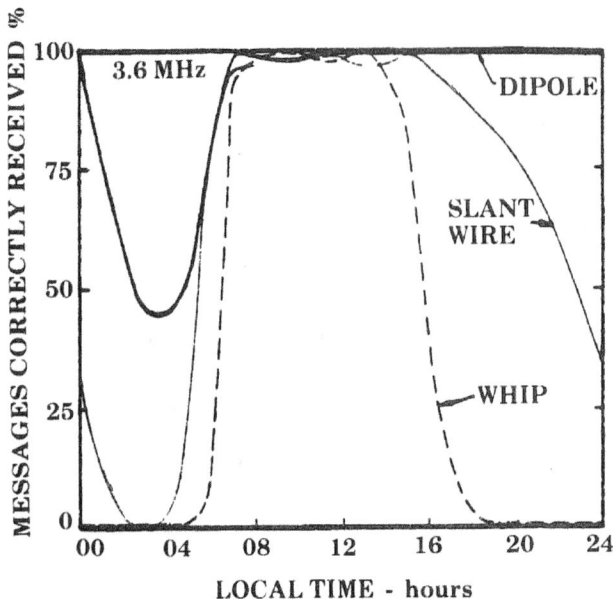

Figure 8. Communication success with AN/PRC-74 as a function of time of day and antenna type over 12-mi path in low mountains, spring and summer 1963.

ground, excellent results can be obtained on both air-to-ground and ground-to-ground paths. During these tests, NVIS radio was used to replace HF groundwave nets and VHF nets that required several ground and airborne retransmission stations to communicate to stations previously unreachable (without relays) due to skip zones and unsuitable terrain.

We urge that those whose problems we have described try NVIS and observe the positive results. We also urge the Army to incorporate NVIS into communication training and amend the reference TMs and FMs to include the use of the NVIS technique.

wave HF techniques being used supported, for the most part, tactical operations. At present, and more importantly after implementation of Division 86 - style operation, HF radio and the NVIS mode take on new importance. HF radio is quickly deployable, securable, and capable of data transmission. Therefore, it will be the first (and frequently the only) means of communicating with fast - moving or farflung units. Also, it may provide the first long-range system to recover from a nuclear attack. The planned Objective HF Radio (OHFR) will meet these requirements.

With this new reliance on HF radio, the communications planners and operators must be familiar with NVIS techniques and their applications and shortcomings in order to provide more reliable and responsive communications for the field commanders. In order to do this, NVIS must be learned, it must be taught and it must be used. Field tests with both the New Hampshire and New Jersey Army National Guard (50th Armored Division) have shown that even using obsolete radio equipment (for example, the *AN/GRC-26D)* and standard wire dipoles *(AN/GRA-* 50) cut to the right frequency and located *1/4* to *1/10* wavelength from the

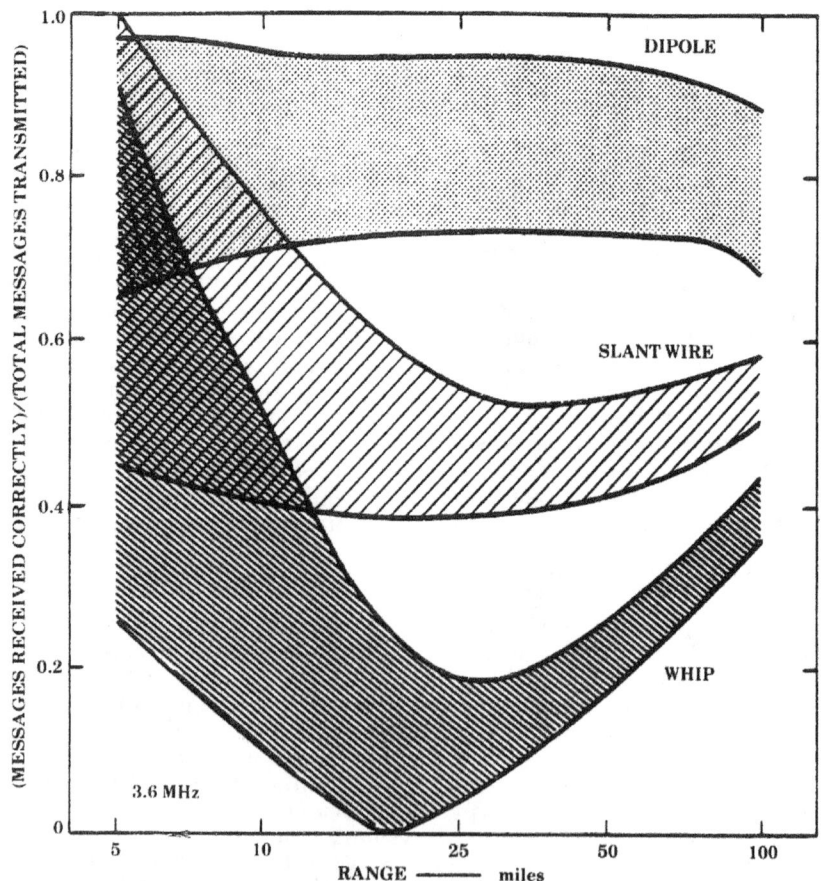

Figure 9. Communication success as a func of range for AN/PRC-74 in mountainous and varied terrain - incl jungle - in Thailand

1. US Army, "Field Radio Techniques," FM 24-18, Washington, D.C., 30 July 1965. (See also coordination draft revision, "Tactical Single-Channel Radio Communications Techniques," 1 September 1977).

2. US Army, "High Frequency Radio Communications," *FM 11-65*, Washington, D.C., 31 October 1978.

3. US Army, "Antennas and Radio Propagation," *TM 11-666*, Washington, D.C., 9 February 1953.

4. K.A. Norton, "The Calculation of Ground-Wave Field Intensity Over a Finitely Conducting Spherical Earth," *Proc, IRE*, Vol. 29, No. 12, pp. 623-639, December 1941.

5. G.H. Hagn, "VHF Radio System Performance Model for Predicting Communications Operational Ranges in Irregular Terrain," *IEEE Trans. COM*, Vol. COM-28, No. 9, pp. 1637-1644, September 1980.

6. K. Davies, *Ionospheric Radio Propagation*, NBS Monograph 80, National Bureau of Standards, Boulder, CO, 1965. (Available from US Government Printing Office, Washington, D.C.).

7. B.C. Tupper and G.H. Hagn, "Nap-of-the-Earth Communication Program for US Army Helicopters," Final Report, Contract DAAB07-76-C-0868, SRI Project 4979, AVRADCOM Technical Report TR-76-0868-F, SRI International, Menlo Park, CA, June 1978. AD-A-063 089.

8. B.V. Ricciardi and J.V. Brune, "Modern HF Communications for Low Flying Aircraft," AGARD-CP-263, NATO Aerospace Research and Development Conference on Special Topics in HF Propagation, Lisbon, Portugal, 28 May-1 June 1979.

9. G.H. Hagn, "On the Relative Response and Absolute Gain Toward the Zenith of HF Field-Expedient Antennas--Measured with an Ionospheric Sounder," *IEEE Trans. Ant. and Prop.*, Vol. AP-21, No. 4, pp. 571-574, July 1973.

10. J.R. Shirley, "The Shirley Aerial - A Vertically Beamed Antenna for Improved Short Distance Sky Waves," Report 8/52, Operational Research Section, Far East Land Forces, Great Britain, 1952.

11. C. Barnes, J.A. Hudick and M.E. Mills, "A Field Guide to Simple HF Dipoles," SRI Project 6183, Contract DA 28-043 AMC-02201(E), Stanford Research Institute, Menlo Park, CA, p. 28, March 1967.

12. G.H. Hagn and J.E. van der Laan, "Measured Relative Responses Toward the Zenith of Short-Whip Antennas on Vehicles at High Frequency," *IEEE Trans. Vehic. Techn.* Vol. VT-19, No. 3, pp. 230-236, August 1970.

13. G.H. Hagn, "Orientation of Linearly Polarized HF Antennas for Short-Path Communication Via the Ionosphere Near the Geomagnetic Equator," Research Memorandum 5 (revised), Contract AMC-00040(E), SRI Project 4240, Stanford Research Institute, Menlo Park, CA, June 1964. AD 480 592.

14. P. Nacaskul, "Orientation Measurement in Thailand with HF Dipole Antennas for Tactical Communications," Special Technical Report 31, Contract DA-36-039 AMC-00040(E), SRI Project 4240, Stanford Research Institute, Menlo Park, CA, June 1967. AD 675 460.

15. G.H. Hagn and G.E. Barker, "Research-Engineering and Support for Tropical Communications," Final Report, Contract DA-36-039 AMC-00040(E), SRI Project 4240, Stanford Research Institute, Menlo Park, CA, February 1970. AD 889 169.

16. G.E. Barker, "Measurement of the Radiation Patterns of Full-Scale HF and VHF Antennas," *IEEE Trans. AP*, Vol. AP-21, No. 4, pp. 538-544, July 1973.

17. C.L. Rufenach and G.H. Hagn, "Predicted Useful Frequency Spectrum of Man-Pack Transceivers Considering Short-Path Skywave Propagation and Comparison with C-2 Sounder Data From Bangkok, Thailand," Special Technical Report 15, Contract DA-36-039 AMC-00040(E), SRI Project 4240, Stanford Research Institute, Menlo Park, CA, August 1966. AD 662 065.

18. Steven R. Holmes, "Near Verticle Incidence Skywave for Military Application," US Army Communications-Electronics Engineering Installation Agency, Fort Huachuca, AZ, August 1979.

19. CCIR, "World Distribution and Characteristics of Atmospheric Radio Noise," Report 322, International Radio Consultative Committee, International Telecommunication Union, Geneva, 1964.

20. CCIR, "Man-Made Noise," Report 258-4, International Radio Consultative Committee, International Telecommunication Union, Geneva, 1982.

21. G.H. Hagn and W.R. Vincent, "Comments on the Performance of Selected Low-Power HF Radio Sets in the Tropics," *IEEE Trans. Vehic. Tech.*, Vol. VT-23, No. 2, pp. 55-58, May 1974.

22. USAF, "High Frequency Radio Communications in a Tactical Environment," AFCSP 100-16, HQ, Air Force Communications Service, Scott AFB, IL 62225, 20 September 1968.

George Hagn received the BSEE and MSEE degrees from Stanford University in 1959 and 1961 respectively. He became a member of the Technical staff of SRI International (formerly called Stanford Research Institute) in 1959. He is currently a Program Director in SRI's Telecommunications Sciences Center in Arlington, VA. Hagn has been involved in studies of tactical communications for over 20 years and has co-authored several books, and contributed to over 100 technical papers and reports. He is a member of the (IEEE), the International Union of Radio Science (URSI), the American Geophysical Union (AGU), the Armed Forces Communications Electronics Association (AFCEA), and the Old Crows.

David M. Fiedler was commissioned in the Signal Corps upon graduation from the Pennsylvania Military College in 1968. He is a graduate of the Signal Officers Basic Course, the Radio and Microwave Systems Engineering Course, the Signal Officers Advanced Course and the Command and General Staff College. Mr. Fiedler has served in Regular Army and National Guard Signal, Infantry, and Armor units in CONUS and Vietnam. He holds degrees in physics and engineering and an advanced degree in Industrial Management. He is an engineer with the US Army Communications Systems Agency, Fort Monmouth, NJ, and is also the Assistant Division C-E Officer of the 50th AD (NJNG).

Skip the "skip zone": we created it and we can eliminate it
by Lt. Col. David M. Fiedler

Current doctrine is wrong. There can be a skip zone if the communicator selects an antenna with too low a radiation angle, but there is no skip zone unless you, the communicator, create it! . . . We must banish forever the term "skip zone" and the thinking that created it.

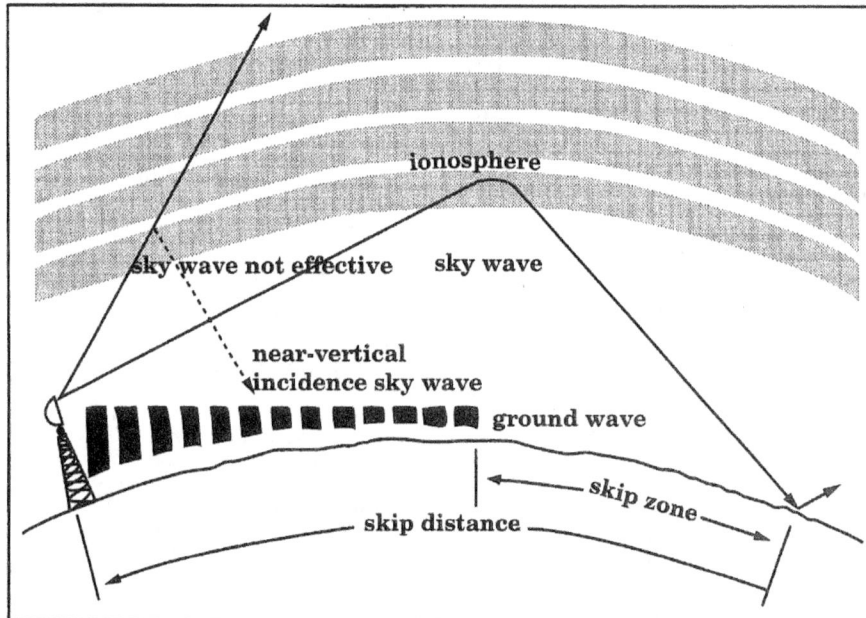

Figure 1. This illustration from FM 24-18 and other publications shows the incorrect concept of a skip zone. If such a skip zone exists when not desired, the communicator has improperly selected the antenna or antenna height.

For many years, Army publications dealing with high frequency (HF) radio communications (i.e. *FM 24-18, FM 24-1, FM 11-65, TM 11-666,* and others) have used the diagram shown in Figure 1. These publications explain how HF radio signals (2-30 MHz) are propagated either as groundwaves or skywaves. They correctly define a groundwave as energy radiating along the surface of the earth until it reaches a point (10 to 70 miles distant) where the energy level becomes too low to be of use for communications. And they define a skywave as energy radiated at an upward angle from the antenna and reflected back to the earth's surface by the ionosphere, adding, however, that this reflection should be expected at a point no less than 100 miles from the antenna. Thus, they claim there is a gap, or "skip zone," of 30 to 90 miles (beginning where the groundwave becomes too weak for communication and ending where the skywave returns to earth) in which HF radio communications are ineffective. To quote *FM 24-18:* "There is an area

called the skip zone in which no useable signal can be received from a given transmitter operating at a given frequency. This area is bounded by the outer edge of useable groundwave propagation and the point nearest the antenna at which the skywave returns to earth." This doctrine is wrong. There *can* be a skip zone if the communicator selects an antenna with too low a radiation angle, *but there is no skip zone unless you, the communicator, create it!*

After many years of urging by myself and others, the December 1984 issue of *FM 24-18* finally included an appendix (Appendix N) on near vertical incidence skywave (NVIS) HF propagation. This appendix clearly shows how, by adjusting antenna heights and transmitter frequencies, an operator can obtain high angle radiation and eliminate skip zones. Apparently, however, the subject area experts of the Signal School did not grasp the significance of Appendix N, since they allowed Figure 1 (which is a reproduction of Figure 2-15 in the manual), an

illustration of a "skip zone," to remain in the same manual untouched. Because the ranges covered by the so-called skip zone are of particular significance to the Army for many tactical reasons, it is important to understand what Appendix N is saying. Using a slightly different approach than that used in Appendix N, I will try to demonstrate again how the "skip zone" can be avoided.

First, however, we need to take a quick look at the accompanying illustrations. Figure 2 shows the relationship between the angle of radiation and the distance covered, assuming an average height of the ionosphere. It shows, for example, that if a station wishes to continuously cover a distance up to 200 miles (approximately the depth and width of an Army corps), it needs to radiate its signal at all angles between 52 and 87 degrees toward the zenith (i.e. in a vertical direction). Figure 3 shows that a horizontal dipole antenna .25 wavelengths above ground will direct most of its energy between these angles. Figure 4 shows that a vertical antenna .5 wavelengths above ground will do likewise with almost equal efficiency. Thus, provided that the operating frequency does not exceed the maximum useable frequency (MUF), either of these simple, commonly available configurations will do the job of directing energy nearly vertically so that it will be scattered and reflected downward by the ionosphere.

MUFs can easily be determined using ionospheric sounding or propagation prediction tables. If these are not available, use the common rule of thumb (2-4 MHz nighttime, 4-8 MHz daytime). The best operating frequency is usually about 20-25 percent below the MUF. Figure 5 is a compilation of all angles of radiation vs. antenna height in wavelengths for horizontal (black curves) or vertical (white curves) antennas. Using this figure, we see the same results as before. A horizontal dipole, .1-.25 wavelengths above ground, radiating energy at all angles between 51 and 87 degrees, falls directly within the first black curve. Similarly, a vertical antenna, .5 wavelengths above the ground, radiating energy at all angles between 50 and 90 degrees, falls directly within the first white curve.

Wavelength (l) can be calculated using the equation:

$$\text{wavelength } (\lambda) = \frac{300}{\text{frequency (in MHz)}}$$
in meters

For example, if our operating frequency were 5 MHz, the wavelength would be 300/5=60 meters. For a horizontal dipole antenna to radiate energy at the necessary angles, it would have to be 6 (.1λ) to 15 (.25λ) meters above the ground. Similarly, a vertical dipole would have to be 30 meters (.5λ) above the ground for the same result at this operating frequency.

Knowing the above information, operators can use the following steps in order to provide continuous, skip zone free HF radio communications for military operations.

1. From the operations order or other directive, determine the ranges to all stations in the net. Stations within 500 miles (well within a division or corps area) can be easily reached using simple dipole antennas-provided a propagating frequency is available.

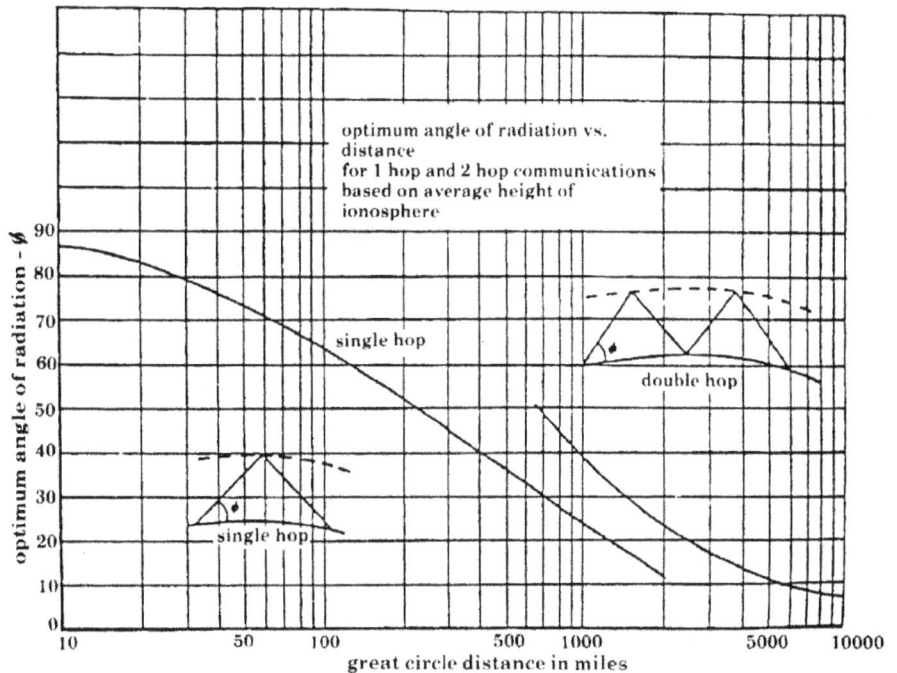

Figure 2. Radiation angle vs. range (from The Rules of the Antenna Game).

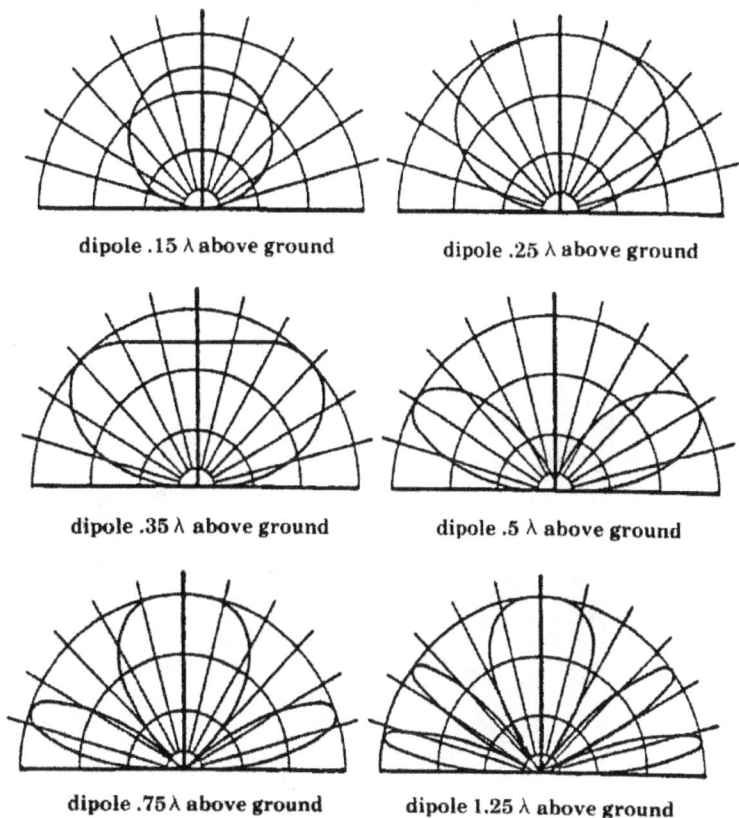

Figure 3. Horizontal dipole radiation patterns at various heights (in wavelengths) above the ground (from Air Force Comm. Pam. 100-16).

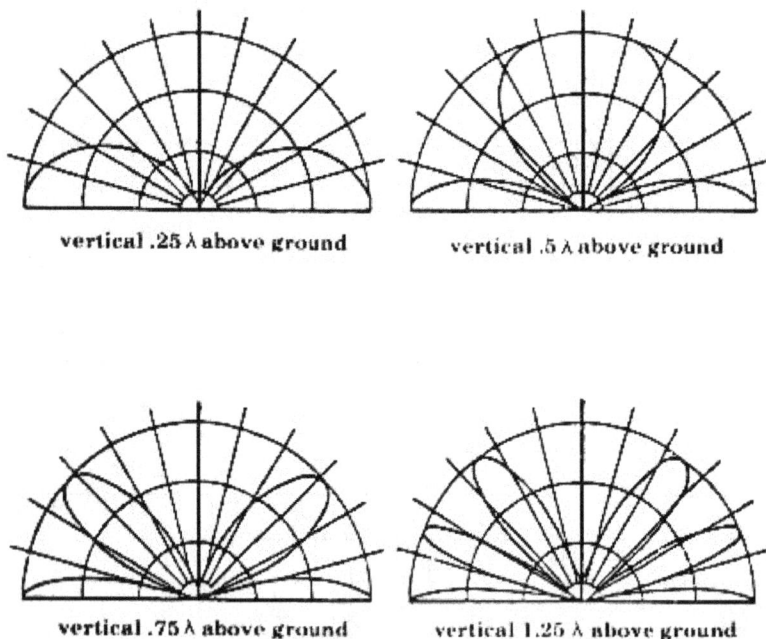

vertical .25 λ above ground   vertical .5 λ above ground

vertical .75 λ above ground   vertical 1.25 λ above ground

Figure 4. Vertical dipole radiation patterns at various heights (in wavelengths) above the ground (from Air Force Comm. Pam. 100-16).

2. From Figure 2, determine the range of angles required to transmit energy over the distance required.

3. From ionospheric sounders, propagation tables, or other predictions, determine the range of frequencies required to support propagation throughout the day. (The frequency will change depending on the time of day, thus requiring frequency shifts and retuning of radios and antennas for 24-hour operation. Typical changes are shown in Figure 6.)

4. From the authorized frequency list provided to your unit, select operating frequencies which fall within the range determined in Step 3. (If no authorized frequencies fall within this range, you will not communicate past groundwave range.)

5. After selecting the operating frequency, select the antenna type desired (vertical dipole, horizontal dipole, etc.).

6. Calculate the required antenna height in wavelengths.

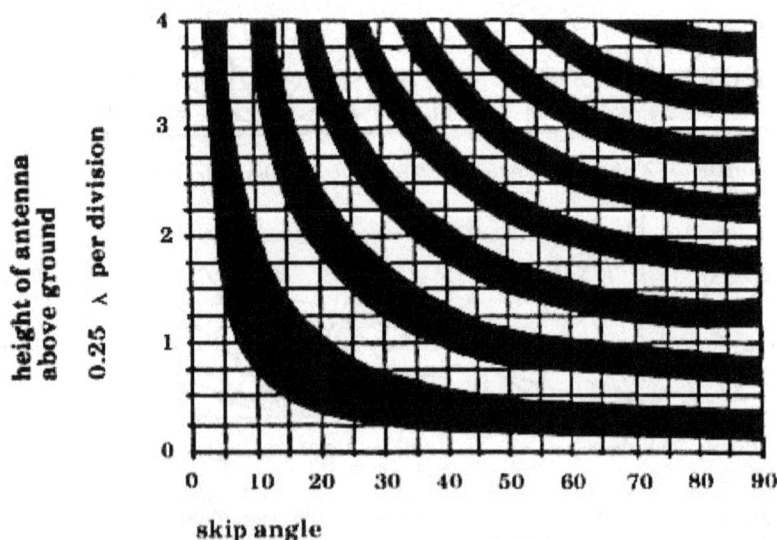

Figure 5. Radiation angle vs. antenna height above ground. Select the antenna height for range of radiation angles desired (from The Rules of the Antenna Game).

$$\text{wavelength in meters} = \frac{300}{\text{frequency in MHz}}$$

7. Mount the antenna at or near this height. (Physical height is not too critical. Height plus or minus .1 wavelengths will function well.)

8. Tune the equipment to the operating frequency and operate.

9. Change frequencies as required by propagation. (Since any antenna height between .1 and .25 wavelengths will direct the energy in a verticle direction, raising and lowering the antenna will almost never be necessary once it is positioned.)

This information in conjunction with Appendix N of *FM 24-18* will allow the tactical communicator to engineer HF systems which will permit 24-hour-a-day operation at ranges up to 500 miles without skip zones. There are two critical elements of this approach which operators must understand. First, propagating frequency bands vary throughout the day depending on the time (sun position). In order to remain in communication, operators must adjust the frequencies, typically at sunup and sundown; however, some propagating frequency will always exist. Second, operators must properly match the antenna impedance to the transmitter in order to efficiently radiate energy. Once this is accomplished, they must select the proper antenna heights in order to direct the radiated energy at the desired angle. Due to the symmetry of high angle radiation, antenna orientation is not a factor and field strength patterns are totally omnidirectional.

The operational experience of the state area command of the New Jersey Army National Guard net shows that 31 stations operating at ranges varying from 5 to 125 miles can provide reliable communications under all conditions if the procedures outlined above are followed and the critical elements of antenna and frequency usage are properly considered.

In light of the above information, let me again urge the Army Signal Center and School to incorporate this information along with the information in *FM 24-18*, Appendix N into the Signal School's program of instruction and to change all Army literature to reflect this information. As we move into the late 1980s and

1990s, it is becoming more and more obvious that HF radio will be the best, if not the only, means of radio communications when beyond line of sight (BLOS) ranges are required, when satellite communications have been destroyed or disrupted, and in situations such as quick recovery after a nuclear exchange, high intensity short duration operations, or anti-terrorist operations.

Both the Air Force and the Navy have recognized these facts and consequently devote many hours of instruction to antennas, radio propagation, frequency selection, and the characteristics of skywave paths in their programs of instruction and manuals. Of particular note is Air Force *Communications Pamphlet 100-16*, which has an excellent chapter on short skywave paths. Unfortunately, in my recent conversation with graduates of the HF Radio Operators Course (31C) and the Communications Electronics Officer Course (25A) (admittedly a small sample, but I think representative), I detected a decided lack of knowledge in this area. Such a lack of knowledge cannot be tolerated if we are to utilize our enormous investment in HF communications equipment effectively and efficiently. The technology exists today. It is up to the Signal Corps (particularly the Signal Center and School) to get this information into the program of instruction and out to the troops where it will do the most good. Above all, we must banish forever the term "skip zone" and the thinking that created it.

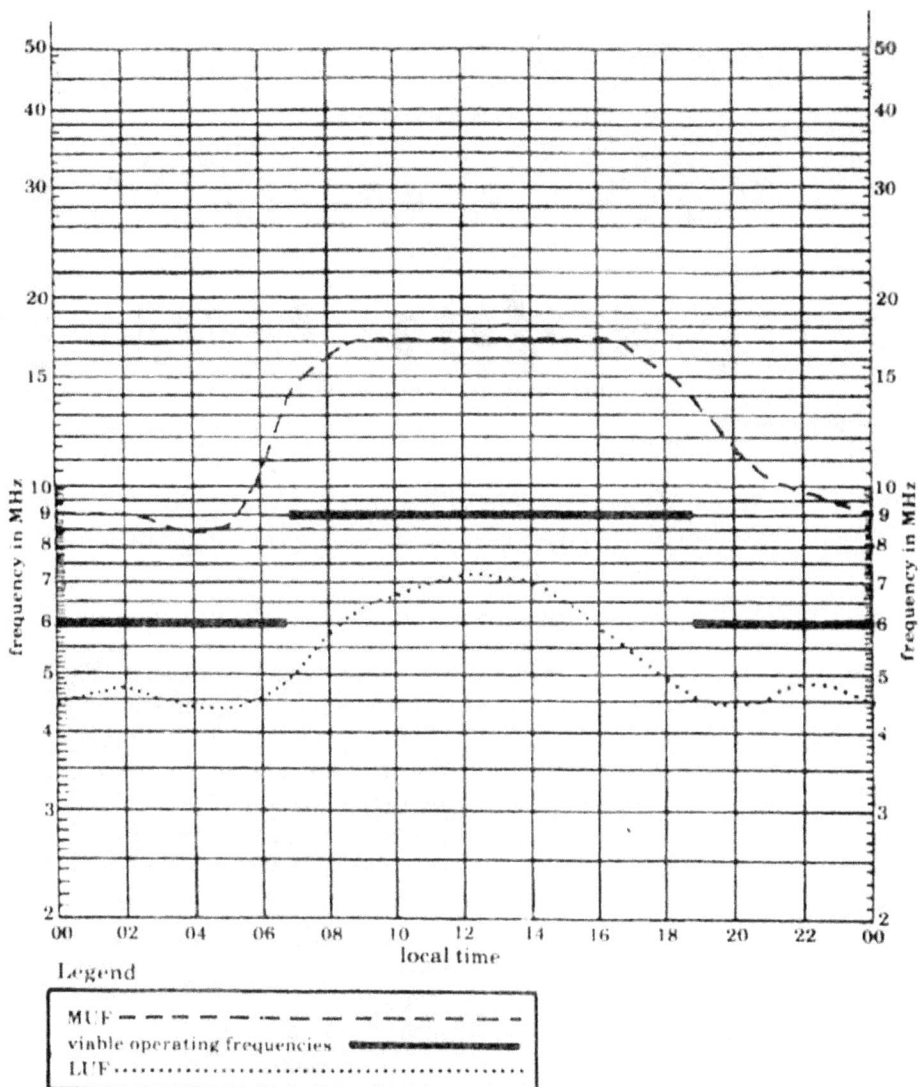

Figure 6. Typical daily variation of the maximum useable frequency (MUF) and lowest useable frequency (LUF), along with suggested operating frequencies. Note that the MUF and LUF values are lowest at night and peak around noon. Operators can maintain 24-hour communications by using a frequency safely between the MUF and LUF. In the example above, 24-hour operations can be achieved by operating near 6 MHz from 1900-0700 and near 9 MHz from 0700-1900 (necessitating only two frequency changes per day).

References

1. *Appendix N*, **FM 24-18, Tactical Single-Channel Radio Communication Techniques,** *13 December 1984, Headquarters, Department of the Army, Washington, D.C.*

2. Air Force **Communications Pamphlet 100-16, High Frequency Radio Communications in a Tactical Environment,** *20 September 1968, Department of the Air Force Washington, D.C.*

3. **The Rules of the Antenna Game,** *1984, W5QJ4 Antenna Products, P.O. Box 334, Melbourne, Fla. 32902-0334.*

4. **ARMY COMMUNICATOR,** *Fall 1983, p. 14, "Beyond line of sight propagation modes and antennae," by David M. Fiedler and George H. Hagan, U.S. Army Signal Center and School, Fort Gordon, Ga. 30905.*

Lt. Col. Fiedler was commissioned in the Signal Corps upon graduation from the Pennsylvania Military College in 1968. He is a graduate of the Signal Officers Basic Course, the Radio and Microwave Systems Engineering Course, the Signal Officers Advanced Course, and the Command and General Staff College. He has served in Regular Army and National Guard Signal, infantry, and armor units in CONUS and Vietnam. He holds degrees in physics and engineering and an advanced degree in industrial management.

Lt. Col. Fiedler is presently employed as the chief of the Fort Monmouth Field Office of the Joint Tactical Fusion Program (JTFP) and as assistant project manager (APM) for Intelligence Digital Message Terminals (IDMT). He is also chief of the C-E Division of the NJ State Area Command (STARC), NJARNG. Prior to coming to the JTFP, Lt. Col. Fiedler served as an engineer with the Army Avionics, EW, and CSTA Laboratories, the Communications Systems Agency (CSA), the PM-MSE, and the PM-SINCGARS. The author of several articles in the fields of tactical communications and electronic warfare, he has served as a consultant to the Army Study Advisory Group (SAG) for theater communications and as a member of the Mobile Subscriber Equipment (MSE) Evaluation Board.

NVIS propagation at low solar flux indices

by Ed Farmer

NVIS coverage has been described as similar to squirting a hose with a spray nozzle straight up thus producing an "umbrella" of rain for a substantial radius around the hose.

Near-Vertical Incidence Sky-wave (NVIS) propagation is a high frequency (HF) radio technique that can provide reliable, skipzone free omnidirectional, even coverage that would be impossible either with VHF (or above) or with HF ground wave propagation. [1,2,3,4,5]. These attributes have encouraged its use in emergency beyond line-of-sight communication systems. [6]

NVIS successes have been substantial [3,4,5] but as sunspot activity continues its current cyclic decline we must resist attaching magical properties to it. It is important not to lose sight of how NVIS propagation works and how it is effected by solar activity.

NVIS propagation

The U.S. Army Signal Corps and LTC David Fiedler of the New Jersey National Guard describe NVIS propagation as resulting from high angle radiation reflecting from the ionosphere [1,2,3,5]. The literature is somewhat vague concerning an exact definition, but U.S. Army FM 24-18 [1] states, "In order to attain a NVIS effect, the energy must be radiated strong enough at angles greater than about 75 or 80 degrees from the horizontal on a frequency that the ionosphere will reflect at that location and time."

A more liberal definition would be, "Skip-zone free, omnidirectional, high frequency, ionospheric propagation." Achieving it is facilitated by antennas that direct most of their radiation at high take-off angles, typically more that 45 degrees (See Figure 1, taken from U.S. Army FM 24-18). It also requires operation below the "critical frequency" so that radiated energy is reflected back to earth, even when directed straight up--operation above critical frequency will always produce a "skip zone," but more on critical frequency later. Together, these conditions produce the gap-free coverage for which the NVIS technique is attractive.

Frequency selection is important in NVIS work. FM 24-18 suggests, "...the useful frequency range varies in accordance with the path length. The shorter the path or higher the angle, the lower the MUF {maximum useable frequency} and the smaller the frequency range. In practice, this limits the NVIS mode of operation to the 2-to-4 MHz range at night and to the 4-to-8 MHz range during the day... These nominal limits will vary with the 11-year sunspot cycle and they will be smaller during sunspot minimums."

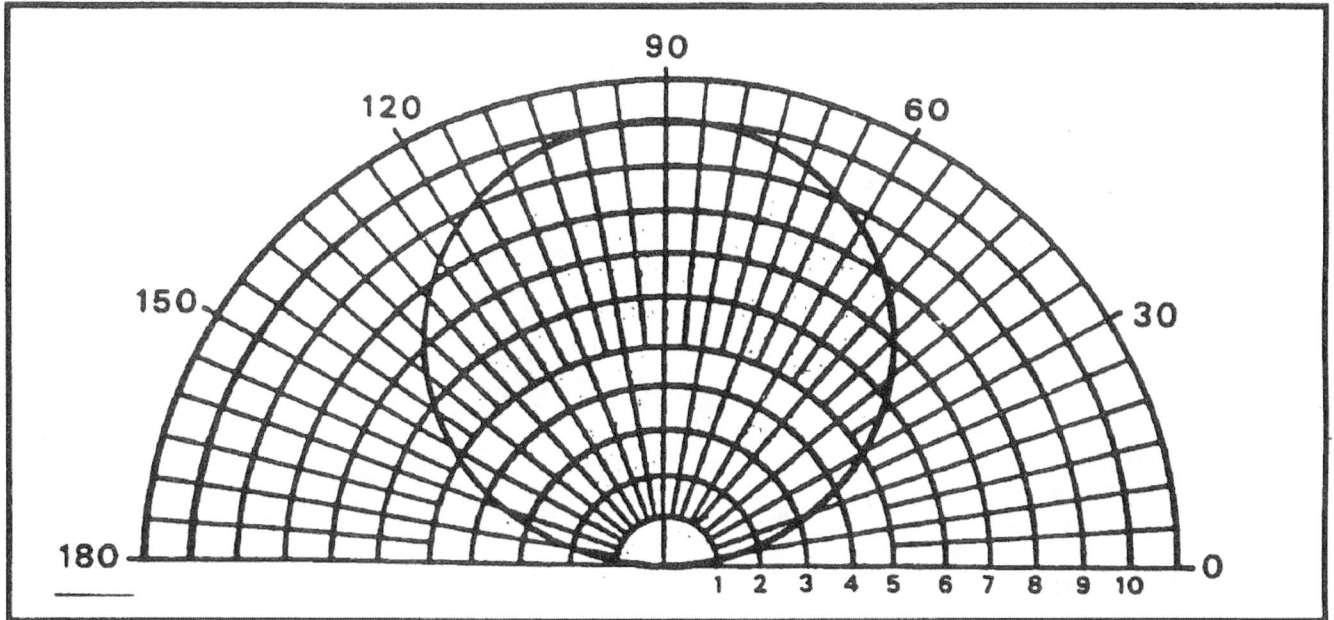

Figure 1 — *The vertical radiation pattern of a NVIS antenna. Most radiation is directed upward. In this example (from Reference [1]) the 3 db points occur at about 45 degrees.*

Practical frequency selection is driven by factors other than where one would prefer to operate. Amateur radio operators have two frequency bands that meet NVIS frequency criteria— the 40 meter (7.0 to 7.3 MHz) and 80 meter (3.5 to 4.0 MHz) bands. The conventional wisdom has been to use 80 meters at night and 40 meters during the day. Military communicators are also faced with frequency selection limitations. Most frequencies between 2 and 4 MHz are allocated to other services.

Frequency selection has not been a problem during the higher phases of the sunspot cycle. In fact, most of the recent testing [4] has occurred during a period of high sunspot activity. Frequency selection and the availability of frequencies close to 2 MHz will be more critical over the next few years as the sunspot cycle goes through its minimum phase.

It will become increasingly important to understand how NVIS propagation works and to assess the factors that effect it in terms of prevailing conditions and mission requirements.

An NVIS path

A NVIS path results from high angle radiation reflecting from a layer (usually the F-layer) of the ionosphere. The radius of coverage will vary from 200 km to as much as 1000 km. Coverage within the active radius will be quite even.

The active area depends on the transmitting and receiving antennas, the operating frequency, and the condition of the ionosphere. The condition of the

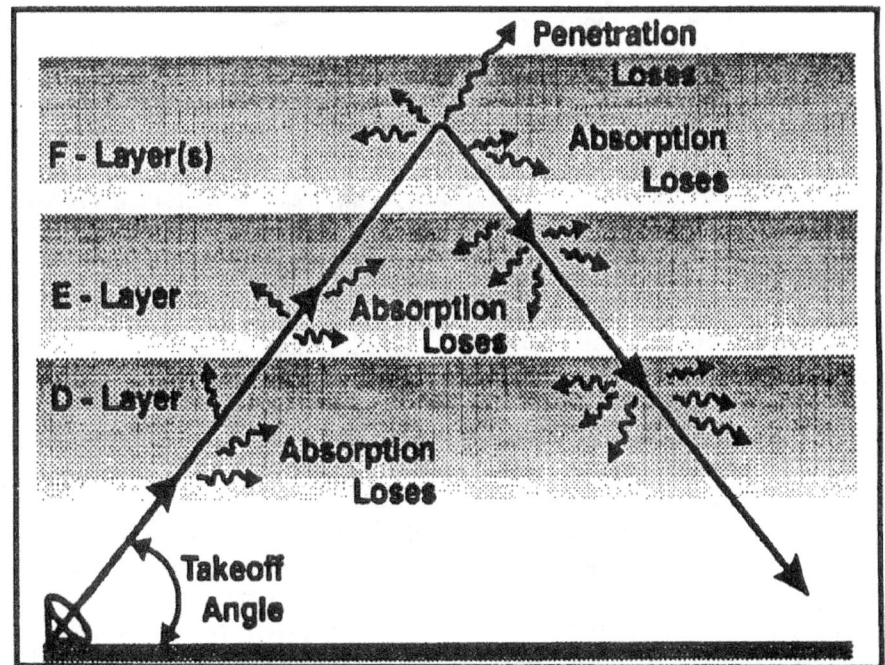

Figure 2 — *NVIS propagation. Energy is radiated at high angle. Some is absorbed by low layers of the ionosphere. Reflection occurs, typically, in the F-layers. Some energy penetrates the F-layers and is lost. (Illustration by SSG(P) Dennis Garman)*

ionosphere depends on the time of day, the time of year, the solar activity (sunspot) cycle, and various transient conditions. Figure 1 shows the vertical radiation pattern typical of a NVIS antenna. Figure 2 illustrates NVIS propagation.

NVIS coverage has been described as similar to squirting a hose with a spray nozzle straight up thus producing an "umbrella" of rain for a substantial radius around the hose. [1]

Selecting frequencies and evaluating communication capabilities for NVIS communication requires a rudimentary understanding of ionospheric propagation.

Ionospheric propagation

Radio waves may travel between a transmitter and a receiver directly, by following the ground (the "ground wave"), or by reflecting from the ionosphere (the "sky wave"). NVIS propagation is sky-wave propagation. On some paths, particularly unobstructed short ones, ground wave propagation may also be possible. These two modes can interact which is usually bad. Good NVIS antennas are typically poor ground wave radiators (and vice versa) so problems are usually easily avoided.

Sky wave propagation results from radio waves reflecting from the ionosphere. Understanding the characteristics of sky wave propagation requires an understanding of the ionosphere.

The ionosphere is a region of charged particles in the upper atmosphere. It results from photoionization by high energy (short wavelength) radiation from the sun. [7] The highest region receives (and absorbs) the most solar energy and therefore becomes more ionized than lower regions.

The more ionized the region, the better it reflects radio waves. Less ionized (lower) regions

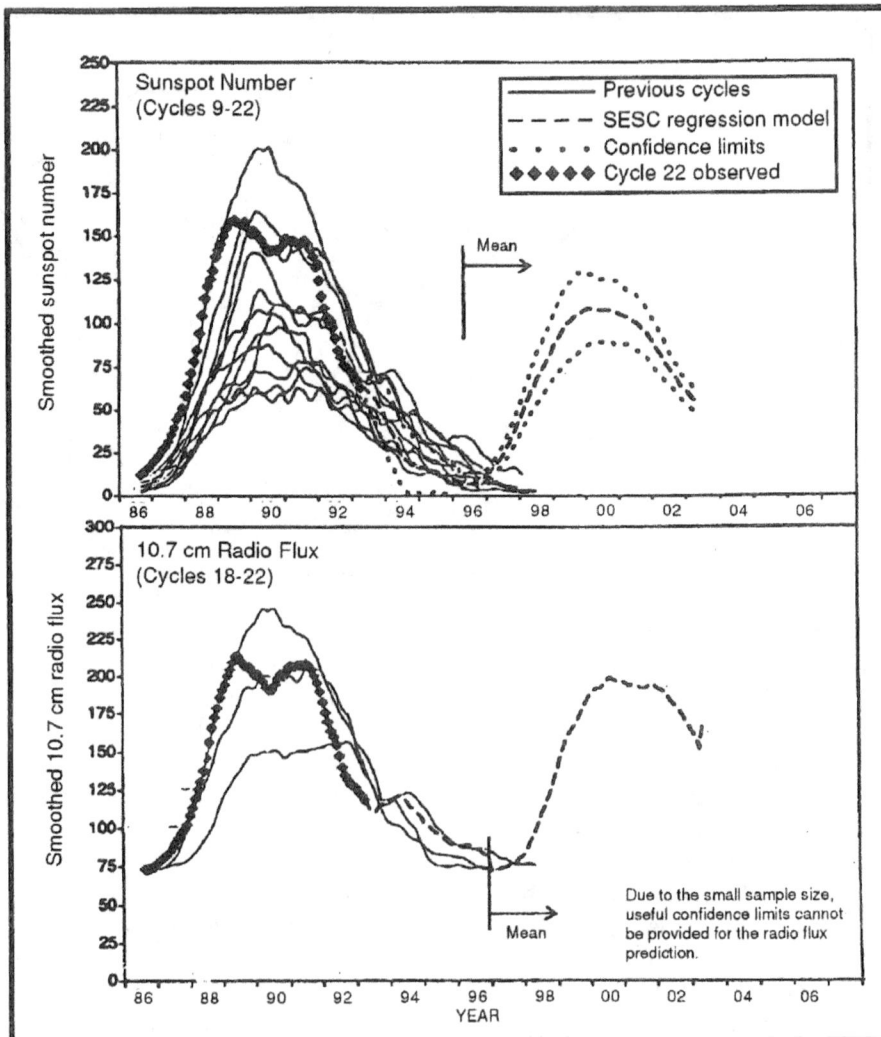

Figure 3 — Solar Activity Cycle (from Reference [4]). Solar activity varies on an 11 year cycle. Sunspot data is available for hundreds of years. As this is being written we are in the waning years of Cycle 22. Note that 10.7 cm Solar Flux correlates well with Sunspot Number. Lt. Col. Fiedler published his landmark paper, "Optimizing Low Power High Frequency Radio Performance for Tactical Operations" in ARMY COMMUNICATOR in the spring of 1989. The Marine Corps experiments that validated Fiedler's work occurred in the spring and summer of 1988, and were reported in the fall 1989 edition of ARMY COMMUNICATOR. During 1988, the Solar Flux Index was increasing rapidly and was between 100 and 200.

absorb radio waves. Lower frequency radio waves are more susceptible to absorption than higher frequencies.

It is productive to think of the ionosphere in terms of "layers." The lowest layer, the D-layer, exists only during daylight hours at heights of about 50 to 90 km. It will not reflect medium or high

frequency waves—it weakens them by absorption. D-layer absorption is always a factor in medium and high frequency daylight propagation. [8]

The E-layer exists at a height of about 90 to 130 km. It's height varies between daylight and dark. It is a factor in medium and high frequency propagation because it can reflect radio waves of sufficiently low frequency. It also attenuates signals that pass through it to the F-layers. [8]

Occasionally, regions of the E-layer will contain cloud-like areas of high ionization. These regions will reflect signals that would normally pass through. Since this effect does not occur with predictable regularity or locality, it is called "Sporadic E." [8]

The F-layer is actually two layers during the day and one layer at night. The F1-layer exists only during daylight at a height of about 175 to 250 km. [8]

The F2-layer is located at a height of about 250 to 400 km. During the day, it is a distinct layer. At night the F1 and F2 layers merge into a single F-layer at a height of about 300 km. [8]

The F2-layer is the principal reflecting region for high frequency communication.

Solar activity and ionospheric propagation

Ionospheric (sky-wave) propagation is the result of radio signals reflecting from the ionosphere. Reflection depends on the operating frequency, the angle at which the radiation encounters the ionosphere, and the degree of ionization.

The degree of ionization depends on excitation from the sun. Obviously there is a significant difference between day and night. There is also a difference that results from solar activity.

There are several measures of solar activity, including the Sun Spot Number, the 10.7 cm (2800 MHz) Solar Flux Index, and the 1.8 Angstrom Background X-ray Flux. These methods can all be related mathematically. For tactical propagation prediction work, the most useful index is the 10.7 cm. Solar Flux Index (SFI), largely because timely values are easy to obtain. [7,9]

Sun spots have been observed for hundreds of years and in 1859 it was determined that they vary on an 11 year cycle. The cycle is assumed to begin at a minimum of solar activity. As this is being written (January of 1994), we are in the 9th year of cycle 22. See Figure 3. [9]

At the peak of the solar cycle there is a very high flux and the ionosphere receives a great deal of excitation. This contributes to its ability to efficiently reflect radio signals. The frequency required to penetrate the ionosphere increases. Absorption in the lower layers also increases but, on balance, higher solar flux is beneficial.

Solar Flux data are readily available which facilitates high and medium frequency propagation prediction. The Solar Flux Index is updated every three hours (beginning at 0000 UTC). It may be obtained by monitoring WWV or WWVH at 18 minutes past each hour. [7]

The Solar Flux Index ranges from a theoretical minimum value of about 65 to more than 250. The higher the number the more active (and the more reflective) will be the ionosphere.

The highest frequency that, when radiated directly upward, will reflect from the ionosphere is called the "critical frequency." Signals above that frequency pass through the ionosphere and out into space. Critical frequency depends only on the condition of the ionosphere—not on any particular communication objective, power level, antenna, or equipment.

Figure 4 (from Reference 10) shows how the configuration of the ionosphere and how the critical frequency are effected by time of day, by time of year, and by solar activity. These figures are typical of four situations. Actual values, of course, depend on the prevailing physical conditions. Figures 4(a) and 4(b) show typical data during a maximum in the solar activity cycle in summer and in winter, respectively. Note that the critical frequency for F-layer propagation is generally above 4 MHz. Figures 4(c) and 4(d) show the same data for the minimum of the solar activity cycle. Note that the F-layer critical frequency is well below 4 MHz at night and only rarely above 6 MHz during daylight. [10]

Frequencies higher than the critical frequency can be useful providing radiation is directed at

Table 1: Data for Cities Used in Path Studies

		Latitude		Longitude		Distance	Take-off
		deg	min	deg	min	km	Angle, deg
Origin:	Sacramento	38	30.7	121	29.5		
Fig.6	Destination						
a.	Stockton	37	53.6	121	14.2	72	82
b.	San Francisco	37	37.1	122	22.4	126	77
c.	Reno, NV	39	29.8	119	46.0	185	71
d.	Fresno	36	46.6	119	43.0	248	65
e.	San Luis Obispo	35	14.2	120	38.4	372	55
f.	Bakersfield	35	19.7	118	59.8	418	51
g.	Lancaster	34	44.4	118	13.0	511	45
h.	Long Beach	33	49.0	118	9.0	602	40
i.	San Diego	32	44.0	117	11.2	751	34

Table 1

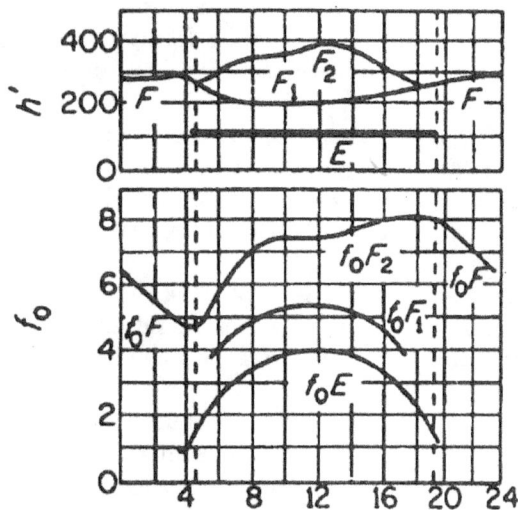

4(a) — Summer during solar
activity maximum.

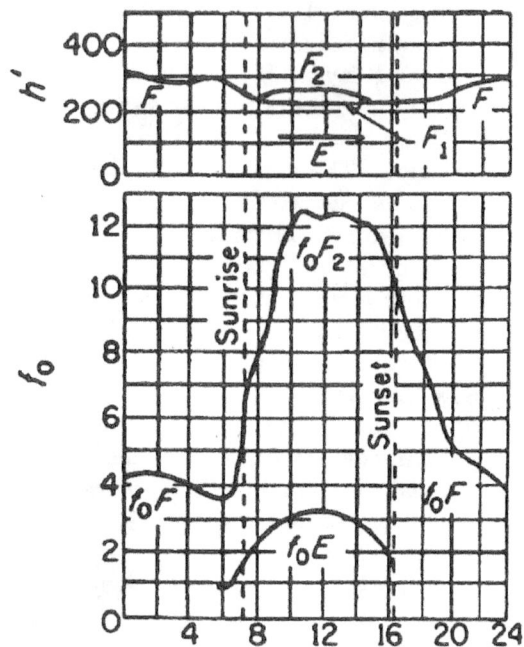

4(b) — Winter during solar
activity maximum.

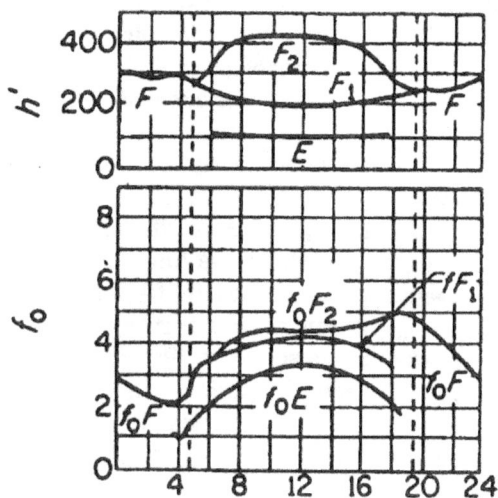

4(c) — Summer during solar
activity minimum.

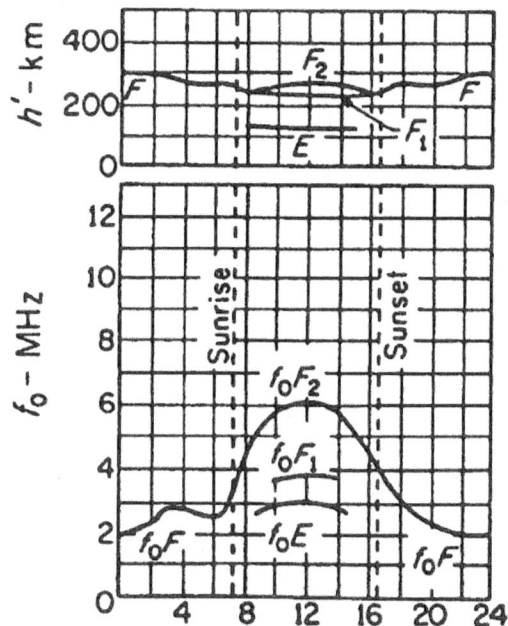

4(d) — Winter during solar
activity minimum

Figure 4 — *Ionosphere configuration and critical frequency showing effects of time of day (local time), time of year, and solar activity cycle, from [10].*

the ionosphere at more grazing angles. Hence, frequencies above the critical frequency can be used for "skip" propagation in which two stations several thousand km apart may be in good communication even though neither station can be heard at most points in between. The signal is said to "skip over" the intermediate listeners. [10, 12]

For any path, however, there is a frequency that will be sufficiently high to penetrate the ionosphere. The highest frequency at which this does not occur more than 50 percent of the time is called the Maximum Useable Frequency, or MUF, for that particular path.

The MUF can be related to the critical frequency (fo) by considering the "takeoff angle." Takeoff angle is the angle between the radiation "ray" of interest and the horizon. The relationship is: MUF = fo / Sin(takeoff angle). [10]

There is a high degree of variability in the ionosphere consequently it is not possible to know the MUF with precision. Generally, the MUF is assumed to be the highest frequency at which the path can be made 50 percent of the time. Consequently, operation near the MUF will not provide a reliable path. As frequency decreases below the MUF a path will exist a higher percentage of the time. The frequency at which the path exists 90 percent of the time is called the "Frequency of Optimum Traffic" or FOT. It will frequently be about 50 to 85 percent of the MUF. [11]

It is also possible to define the lowest frequency at which communication over a given path at a specified power level can take place. In this case, making the path depends on the ability of the communication system to overcome ionospheric absorption and ambient noise. The lowest frequency at which a path can be made is called the Lowest Useable Frequency or LUF. [8, 11]

It is useful to review the various frequencies used to characterize ionospheric propagation.

FO critical frequency: The highest frequency that will not penetrate the ionosphere more than 50 percent of the time when radiation is directed at a takeoff angle of 90 degrees. Critical frequency does not assume any specific communication objective.

Highest probable frequency (HPF): The highest frequency at which ionospheric propagation between specific locations will be available 10 percent of the time.

Maximum useable frequency (MUF): The highest frequency at which ionospheric propagation between specific locations will be available 50 percent of the time.

Frequency of optimum traffic (FOT): The highest frequency at which ionospheric propagation between specific locations will be available 90 percent of the time.

Lowest useable frequency (LUF): The lowest frequency at which ionospheric propagation between specific locations and using specific power levels, receiving equipment, and antennas will be available 50 percent of the time. The LUF can be improved (lowered) by increasing power, using higher gain antennas, or substituting higher performance receiving equipment.

An analytical look at some NVIS paths

Let's look at several NVIS paths originating in Sacramento, California (See Table 1). These paths were selected because they could be significant during certain anticipated natural disasters but also serve to illustrate how NVIS paths of different lengths are effected by SFI.

Propagation data are always specific to the time of year, time of day, and solar activity. In order to illustrate a more-or-less worst case condition, the time of year was selected as the winter solstice (the shortest day of the year), December 21, 1993. The Solar

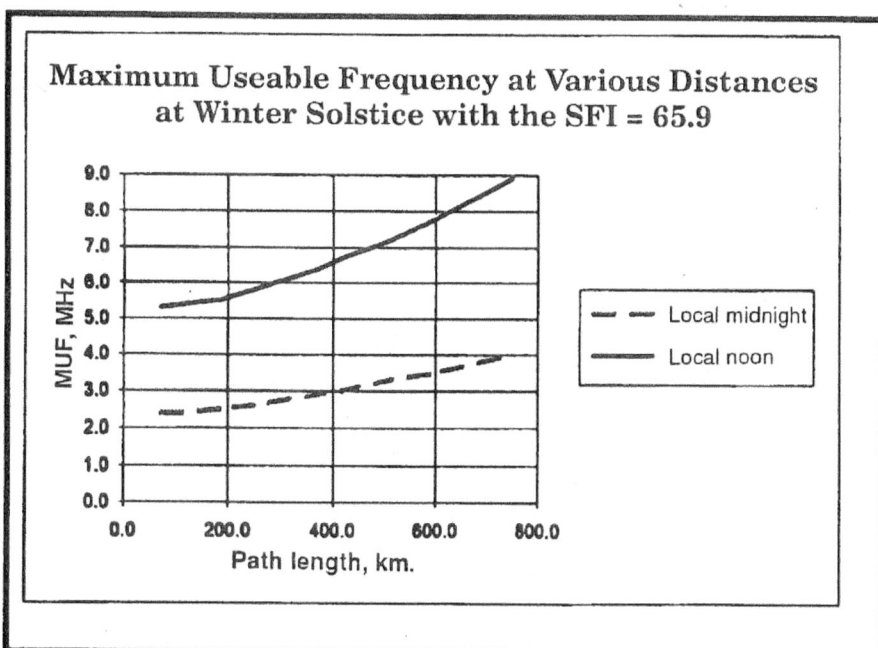

Figure 5 — *Maximum useable frequency increases as path length increases. This graph was prepared for the paths in Table 1 from the propagation data shown in Figure 6.*

Sacramento-Stockton MUF
at Winter Solstice with SFI = 65.9

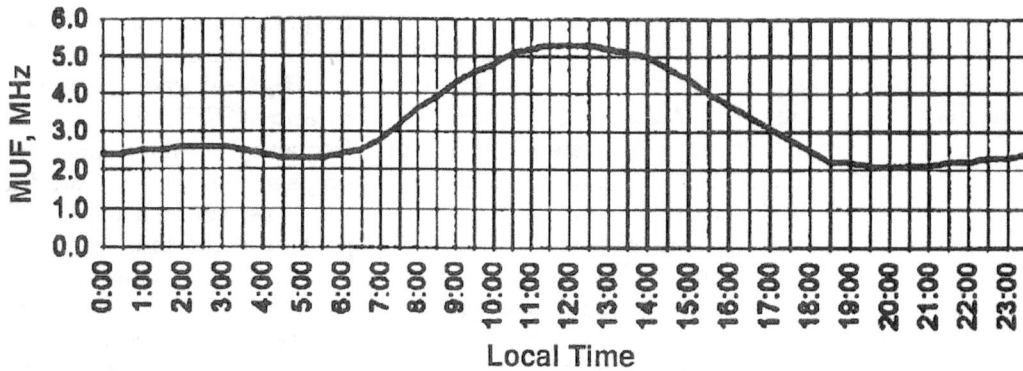

*Figure
6(a)*

Sacramento–San Francisco MUF
at Winter Solstice with SFI = 65.9

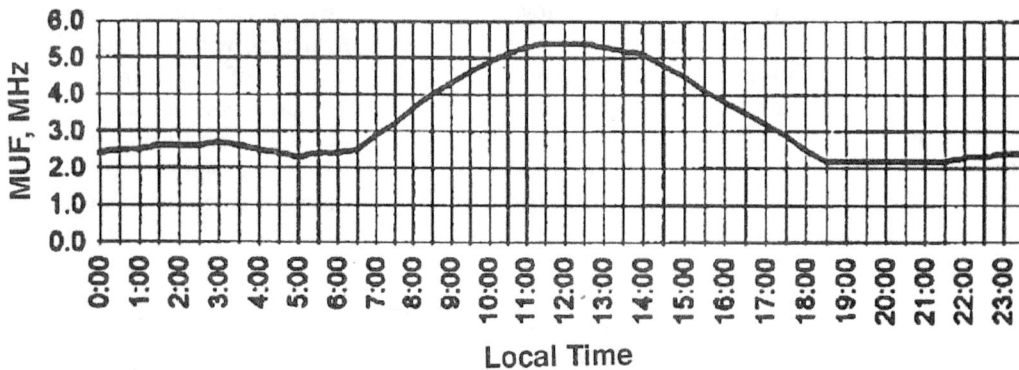

*Figure
6(b)*

Sacramento–Reno MUF
at Winter Solstice with SFI = 65.9

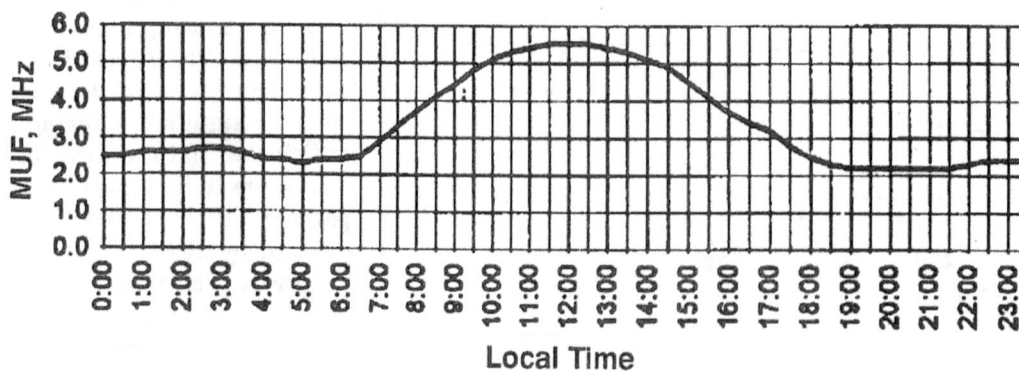

*Figure
6(c)*

Figure 6 (a-i) — Maximum Useable Frequency (MUF) for various paths. Note that the shorter paths have significantly lower MUFs than the longer ones. Terminal and path length data are shown in Table 1.

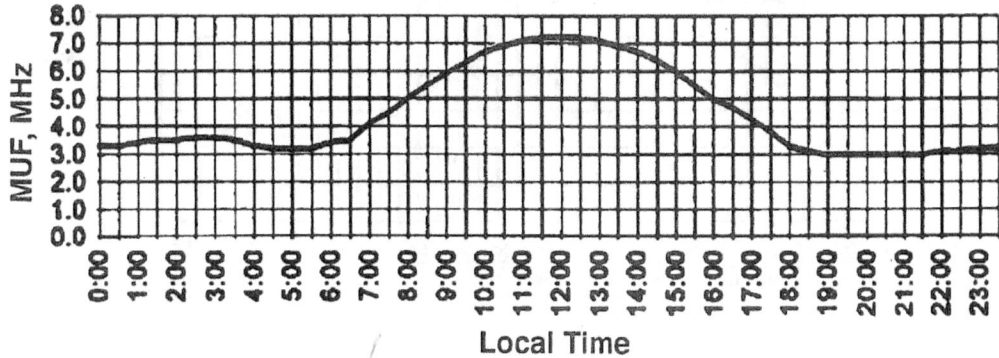

**Sacramento–Lancaster MUF
at Winter Solstice with SFI = 65.9**

*Figure
6(g)*

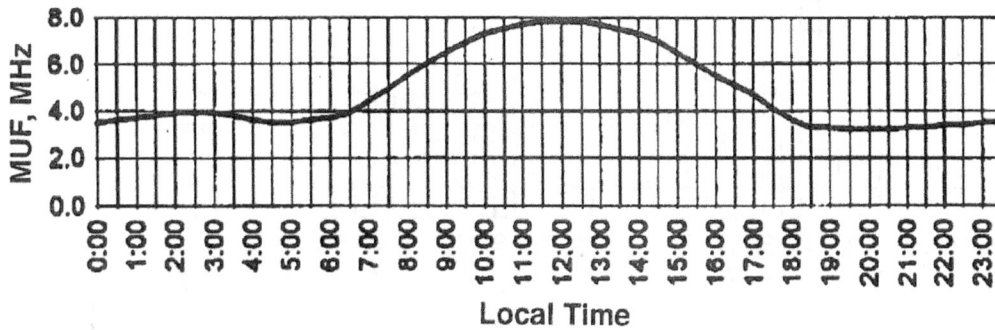

**Sacramento–Long Beach MUF
at Winter Solstice with SFI = 65.9**

*Figure
6(h)*

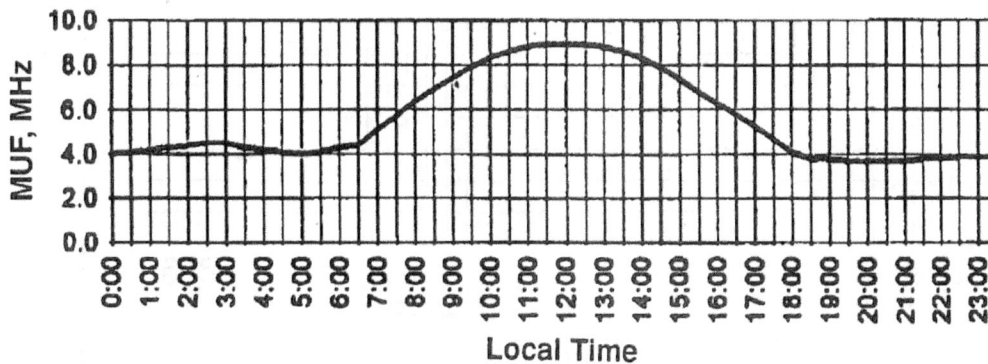

**Sacramento–San Diego MUF
at Winter Solstice with SFI = 65.9**

*Figure
6(i)*

Sacramento–Fresno MUF
at Winter Solstice with SFI = 65.9

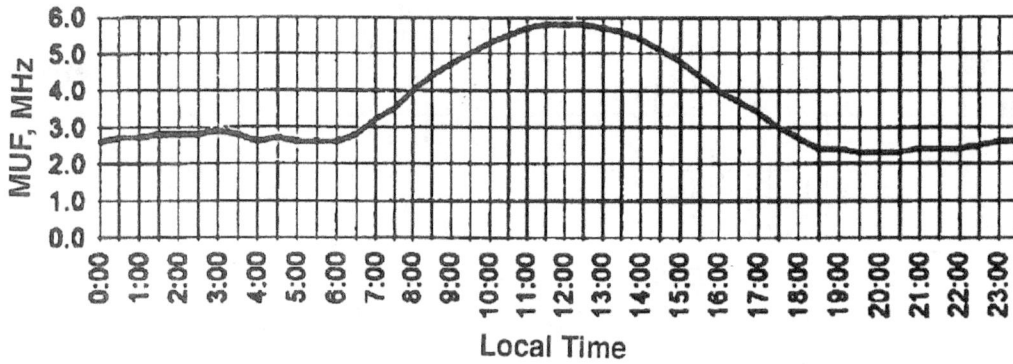

Figure 6(d)

Sacramento–San Luis Obispo MUF
at Winter Solstice with SFI = 65.9

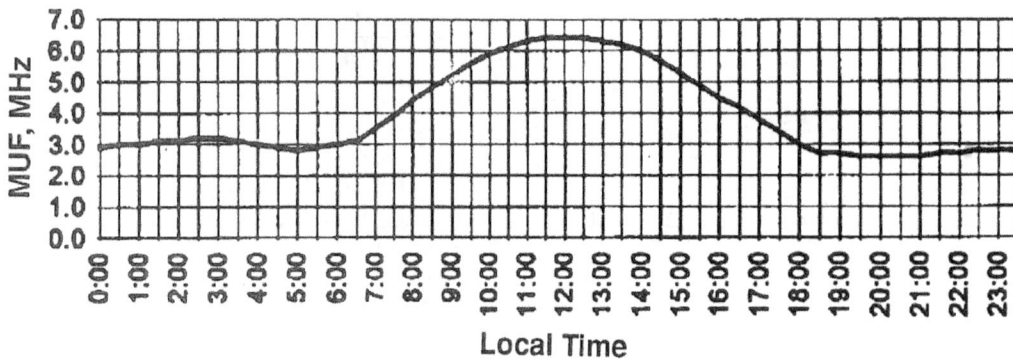

Figure 6(e)

Sacramento–Bakersfield MUF
at Winter Solstice with SFI = 65.9

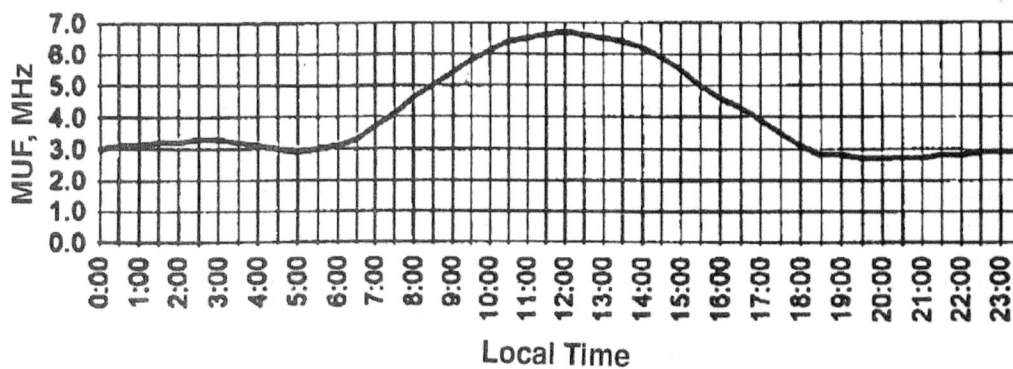

Figure 6(f)

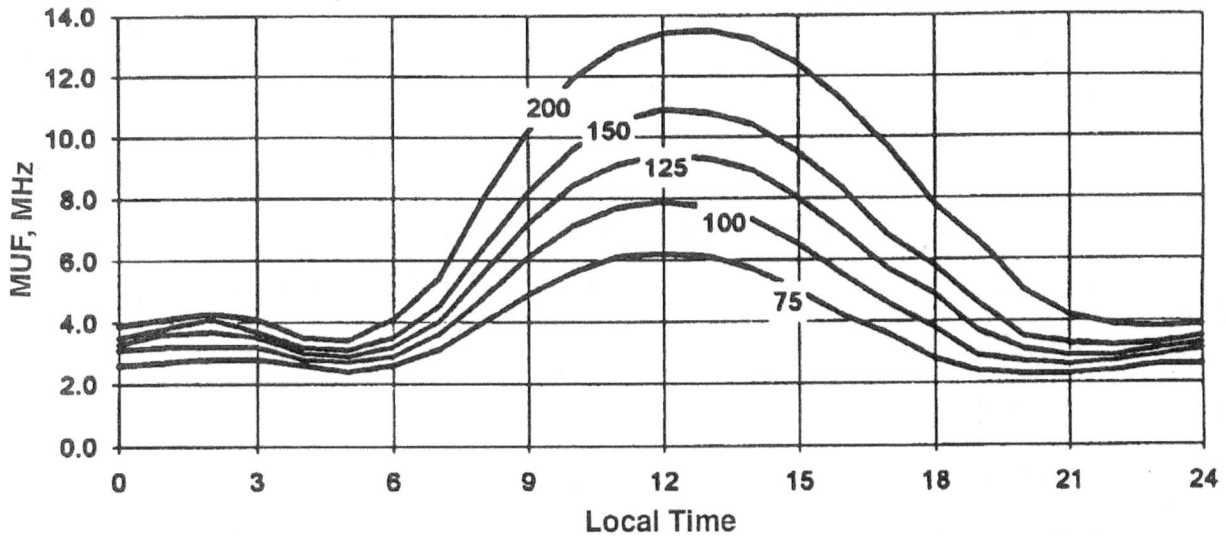

**Maximum Useable Frequency
Sacramento–Reno at Winter Solstice
at Solar Flux Index Indicated**

Figure 7 — Maximum Useable Frequency for a Sacramento - Reno, Nevada path at the winter solstice at the Solar Flux Indices shown.

Flux Index was selected as the minimum value the propagation program that was used (MINIPROP [11]) would accept which turned out to be 65.9. Data were obtained for every half hour.

Figure 5 illustrates the effect of takeoff angle on maximum useable frequency. As the path becomes shorter, the MUF becomes lower. This is because the takeoff angle for shorter paths approaches the vertical, hence the MUF approaches the critical frequency.

Figure 6 illustrates the MUF for the 9 paths shown in Table 1.

Clearly, the greatest challenge for NVIS propagation during low SFI is the shorter paths. On paths under 500 km the predicted MUF is always above 2 MHz but is frequently *not above* 3 MHz. Significantly, the MUF does not rise above 7 MHz for these paths. In fact, it rises above 7 MHz only on the longer paths and then only

for a brief period during the middle of the day.

In terms of amateur radio bands, 40 meters (7.0 to 7.3 MHz) will not be useful for NVIS coverage. Although a path will probably exist to the more distant locations during a short part of the day, the closer locations will be in the "skip zone." The skywave will skip over them rendering communication unlikely.

The 80 meter band (3.5 to 4.0 MHz) will be useful for all locations during daylight hours but will not be reliable at night.

NVIS doctrine [1] would suggest we select a lower frequency for night operations. There are, however, no suitable frequencies available to amateur radio operators. This requires consideration of the medium frequency 160 meter band (1.8 to 2.0 MHz). Millitary systems with more flexibility in frequency assignments are, of course, less affective.

Medium frequency utilization

The 160 meter band will be useful for all paths at night. This, however, is not as good as it sounds. The LUF for a given communication system and objective can be quite high. Ionospheric absorption is a larger factor at medium frequencies (MF) than at high frequency. It is also harder to radiate a strong signal on this band than it is on those of shorter wavelength. [12] It is, however, the only way to make NVIS propagation useable.

It is important that tactical and emergency communicators develop strategies for implementing medium frequency (160 meter) NVIS paths. There are several challenges. The most important ones are equipment and antennas.

Most military HF equipment will not operate below 2 or 3 MHz. While operation at 2 MHz satisfies the physics of the problem, the limited frequency allocations gets in the way. Most modern amateur radio equipment will operate in the 160 meter band.

Full-size resonant antennas for 160 meters are physically large. A half-wave dipole cut for 1.9 MHz would be about 246 feet long. To meet the usual NVIS height objective of 0.1 to 0.25 wavelengths [1,2,3,5] it would have to mounted 49 to 123 feet above the ground. Antennas of this size and mounted at these heights are uncommon, are difficult to erect, and are susceptible to damage during emergency conditions.

Many reduced size antennas exhibit lower radiation resistance. Consequently, resistive losses (e.g., wire and connection resistance) must be minimized or efficiency will suffer.

Even when a good MF skywave is radiated, ionospheric absorption is a significant problem. The sky wave for the AM broadcast band (0.53 to 1.6 MHz) is usually completely absorbed.

During daylight the 160 meter skywave is almost completely absorbed in the D-layer. [12]

Noise is more of a problem at medium frequency. Atmospheric noise is of higher intensity than at HF and man-made noise is more common. These combine to raise the noise floor. [13, 14]

Noise is a complex subject and not conducive to treatment by simple assumptions. To provide some assessment of the approximate magnitude of the effect:

One should expect a 5 to 8 db increase in noise level when frequency is changed from the 40 meter amateur band (7.0 to 7.3 MHz) to the 80 meter band (3.5 to 4.0 MHz). [14]

One should expect an 8 to 9 db increase in noise level when the frequency is changed from the 80 meter band to the 160 meter band (1.8 to 2.0 MHz). [14]

Because of these problems; low antenna gain, low antenna efficiency, higher path losses, and

a higher noise floor; MF paths will require significantly more power than one used to HF operation might estimate.

MUF at higher solar flux index

As the solar flux index increases, the MUF at any time of day also increases. Figure 7 shows the MUF for five values of the solar flux index greater than the minimal value used in preparation of Figure 6. The assumed path was Sacramento - Reno, Nevada. It was selected because it is a medium length path for which ground wave propagation is out of the question. Shorter paths (e.g., Sacramento - Stockton) would have lower MUFs and longer paths (e.g., Sacramento - San Diego) would have higher MUFs.

MUF affected by time of year

Figure 8, prepared using the same solar flux values as Figure 7, illustrates the effect of time of year. On the longest day of the

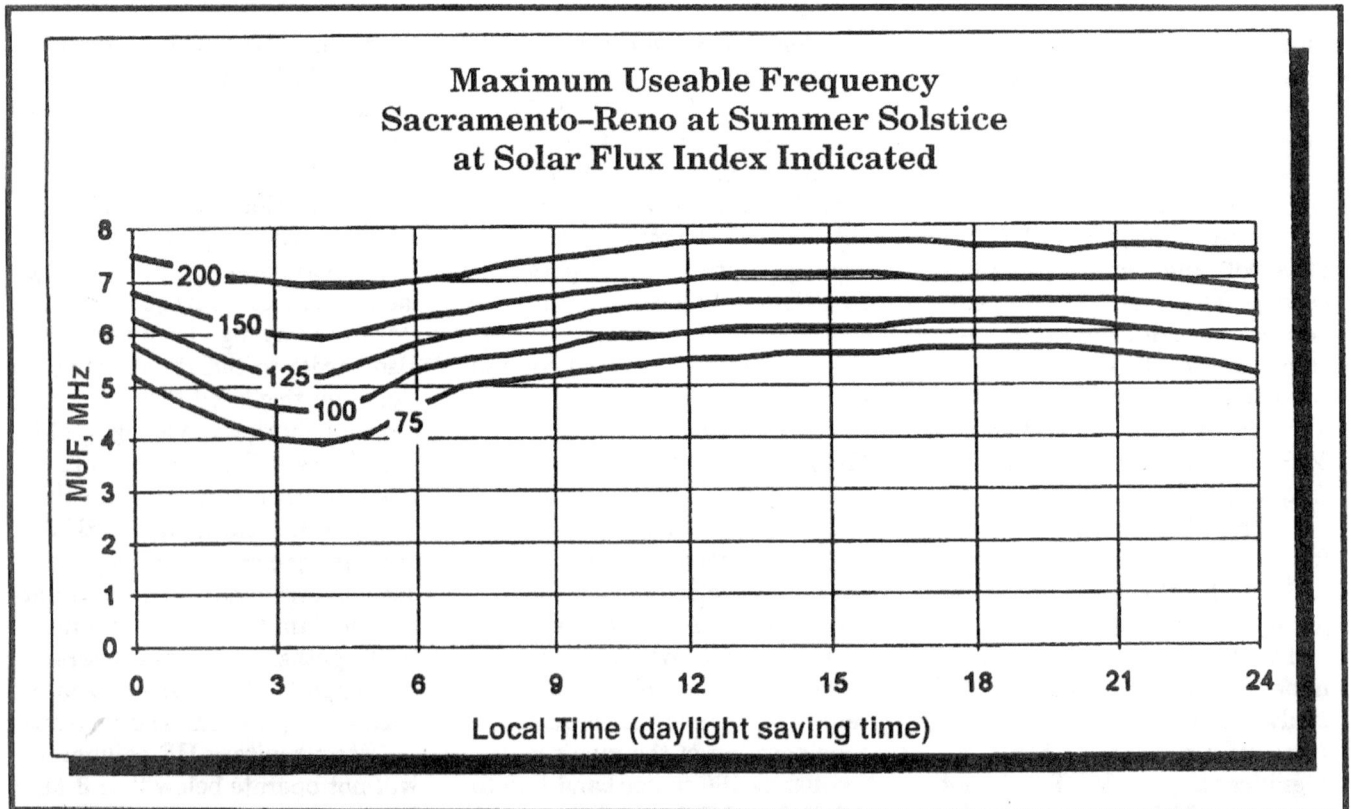

Figure 8 — Maximum Useable Frequency for a Sacramento - Reno, Nevada path at the summer solstice at the Solar Flux Indices shown.

year, the MUF curves show smaller day-to-night variations and the maximum value of the MUF is lower. Most importantly, the minimum value of the MUF is higher. For NVIS communication, winter conditions are clearly more critical than summer.

For the more modern military and commercial systems used in emergency management, the problems of constantly changing path conditions, particularly during low solar activity, have been significantly reduced by the use of automatic link establishment (ALE). HF-ALE systems can deal with HF propagation variables in real time and find the best operating frequency for the conditions.

ALE scans and tests authorized frequencies for a particular path or net until it finds a frequency that will support communications over the path. Each radio in an ALE net constantly broadcasts a sounding signal and "listens" for other sounding signals generated by other net members.

An analysis of these signals by an on-board processor determines the best frequency for communications, and this frequency is then selected automatically for operations.

In summary

There are special challenges for NVIS communication systems during periods of low solar activity. Propagation parameters (e.g., solar flux index) must be carefully considered. The use of propagation prediction tools becomes more important. Frequency selection becomes more critical, especially for nighttime operations.

The use of medium frequencies (e.g., the 1.8 to 2.0 MHz amateur radio band) is a critical element in communications planning. Effort should be devoted to developing easy to deploy antennas and matching systems for this band.

NVIS propagation works, even during periods of low solar activity, but careful consideration of propagation becomes more important as sunspot numbers go through the low portions of their 11-year cycle.

References

[1] U.S. Army Field Manual FM 24-18; **Field Radio Techniques**; Appendix N; December 1984.

[2] Fielder, David M. and Hagan, George; "Beyond line-of-sight propagation modes and antennas"; **ARMY COMMUNICATOR** Magazine; Fall, 1983; Ft. Gordon, GA.

[3] Fiedler, David M.; "Mobile NVIS: the New Jersey Army National Guard approach"; **ARMY COMMUNICATOR** Magazine; Fall, 1987; Ft. Gordon, GA.

[4] Fiedler, David M.; "Marine tests prove Fiedler's NVIS conclusions"; **ARMY COMMUNICATOR** Magazine; Fall, 1989; Ft. Gordon, GA.

[5] Fiedler, David M.; "Optimizing low power high frequency radio performance for tactical operations"; **ARMY COMMUNICATOR** Magazine; Spring 1989, Ft. Gordon, GA.

[6] Harter, Stanly E.; "A Cloud Warmer Antenna is Best for Local H-F Coverage"; State of California Governor's Office of Emergency Service; November 27, 1990.

[7] Rosenthal and Hirman; **A Radio Frequency User's Guide to the Space Environment Services Center Geophysical Alert Broadcasts**; 1990; National Oceanic and Atmospheric Administration Environmental Research Laboratories; Space Environment Laboratory; Boulder, Colorado.

[8] **Reference Data for Radio Engineers**; 6th Edition; Chapter 28, "Electromagnetic-wave Propagation"; Howard W. Sams & Co.; 1977; New York.

[9] Brown, Bob; "Propagation", pp. 35-37; **Worldradio** Magazine; February 1994.

[10] Jordan and Balmain; **Electromagnetic Waves and Radiating Systems**; 2nd Edition; 1968; Prentice-Hall, Inc.; Englewood Cliffs, New Jersey.

[11] Shallon, Sheldon C., W6EL; **MINIPROP**, Version 3; Computer program and User's Manual; 1988; Los Angeles.

[12] **The ARRL Antenna Handbook**; 15th Edition; pp. 23-11 to 23-21 and pp. 4-19 to 4-24; The American Radio Relay League; Newington, CT.

[13] **Reference Data for Radio Engineers**; 6th Edition; Chapter 29, "Radio Noise and Interference"; Howard W. Sams & Co.; 1977; New York.

[14] Handwerker, J., W1FM; **IONSOUND PRO**, computer program and user manual, Version 1.1; 1993; Lexington, MA.

Mr. Farmer, a professional engineer, is president of EFA Technologies, Inc. The former signal soldier has a BS in electrical engineering and an MS in physics, both from California State. He has published over forty articles and two books. He also holds two U.S. patents.

NVIS antenna fundamentals

by Edward J. Farmer, P.E.

This is the second of three articles on NVIS communication. The first, "NVIS Propagation at Low Solar Flux Indices" appeared in the Spring 1994 issue of __Army Communicator__. The third article will be on field antenna experiments. This article deals with design, operation, and understanding of NVIS antennas.

Near vertical incidence sky wave (NVIS) communication involves paths in which radio energy directed upward at high angles is reflected from the ionosphere. Properly done, it provides uniform, dependable communication over corps size (and larger) areas. This is often taken to mean areas with a radius of 250 km or so. [1]

NVIS communication systems require that two issues be properly addressed. First, it is essential that the frequency of operation be selected with full consideration of the prevailing propagation conditions. This is addressed in Reference [2].

Second, it is essential to use an antenna that has a substantial radiation field in the vertical direction. In tactical NVIS applications, selecting an antenna is driven by the specific mission requirements, the field conditions encountered, the transmitter power available, the quality of the base stations with which communication must be maintained, the time and resources available for erection, the prospect of the enemy seeing the antenna or intercepting the signals, the required range of operating frequencies, and the available antenna system components.

The antenna system begins at the radio and involves the antenna tuning unit (ATU), the feed line (transmission line) between the ATU and the antenna, and the antenna itself (see Figure 1). Each element is important in achieving proper operation. They will be discussed in that order.

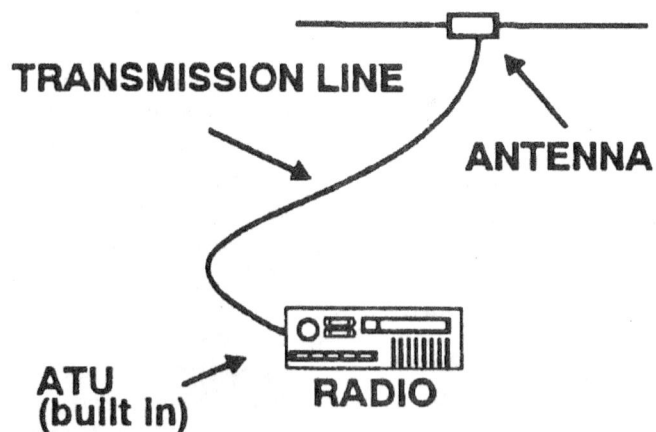

Figure 1. Components of a transmission system including the radio with the ATU, transmission line, and antenna.

ANTENNA TUNING UNIT

The output circuit of modern (e.g., solid state) radios is a moderate impedance—usually 50 ohms. Coupling these radios to any practical transmission line (and hence, antenna) requires matching the impedance presented at the radio's end of the transmission line. This device is an ATU, antenna coupler, antenna matching unit, or transmatch.

This device provides the reactance necessary to compensate for the impedance that appears on the radio end of the transmission line. This impedance depends on the operating frequency, the design of the antenna, its mounting position, the type of transmission line, and its length. This impedance can vary from moderate values of pure resistance (e.g., 72 ohms for a half-wave resonant dipole mounted a half wavelength above ground) to thousands of ohms of reactance coupled with very low (or very high) values of resistance.

Most military radios have an ATU built in so a resonant antenna is not required. For radios such as the AN/PRC-132 that do not have an ATU, a resonant antenna is essential.

The process of "matching" amounts to creating a one-way RF "mirror." [4] RF energy from the transmitter passes through the ATU (in the "transparent" direction of the one-way mirror) into the feed line. It travels down the feed line to the antenna. If the characteristic impedance of the transmission line does not match the feed point impedance of the antenna, some of the energy is reflected back up the transmission line toward the transmitter. The interaction of the incident and reflected energy forms a "standing wave" which is a measurement called the standing wave ratio (SWR). If the feed line is perfectly matched to the antenna (e.g., an antenna with a 72 ohm impedance is fed by a 72 ohm coaxial cable), the SWR is 1:1 or simply "1." If there is an impedance mismatch, the SWR is greater than one perhaps by many times.

Energy traveling back up the transmission line runs into the ATU. When properly adjusted, the ATU re-reflects the energy back down the line toward the antenna. This is the mirror side of the one-way mirror. No signal passes back through the ATU to the transmitter itself.

All the energy supplied from the transmitter through the ATU into the feed line will eventually either be radiated or absorbed in antenna and feed line losses. Some, hopefully most, will be absorbed in the radiation resistance of the antenna and thus will radiate useful signals. Some will be absorbed in the loss resistance of the antenna. Some, often quite a bit, will be absorbed in the transmission line. As long as the ATU is correctly adjusted none will be dissipated in the transmitter itself.

The ATU may be manual or automatic. There are many ways to arrange its reactances [3] however, no matter how implemented, its purpose remains the same.

TRANSMISSION LINES

A feed line or transmission line connects the radio to the antenna's feed point. It may be coaxial cable, parallel wire line ("ladder line" or "twin lead"), or a single wire that may be part of the antenna itself.

A **single wire** feeder is typically part of a single wire (marconi-type) antenna. For example, it is the vertical wire in an Inverted L (a very useful NVIS antenna). Since this feed-wire is part of the antenna, it radiates. Since it is vertical, it radiates mostly in the horizontal direction. This can be useful in situations where ground wave communication is beneficial but it contributes little to the NVIS effect. The amount of radiation from this vertical wire depends on the electrical length of the entire antenna and is greatest when the total length is an odd number of quarter-wavelengths. It can represent a "loss" even though it is, in fact, radiating.

Coaxial cable is the most common transmission line in military systems. It also has the highest loss of the available transmission lines.

Even when both the antenna and the transmitter are perfectly matched to the cable, some of the signal passing through it is absorbed. This loss depends on the design of the cable, its length, and the frequency of operation. At frequencies below 8 MHz this loss is not substantial. For a 100-foot foam dielectric RG-8 coaxial cable, the loss is less than 0.5 dB at 8 MHz and a bit more than 0.2 dB at 2 MHz. To put this in perspective, if 100 watts at 2 MHz entered this cable, about 95.5 watts would enter the antenna.

While it is usually possible to match the ATU to the coaxial cable it is generally not possible to match the antenna, at least not over a wide range of frequencies. Consequently, the mismatch at the antenna causes reflections which create standing waves. As the SWR increases so does loss in the coaxial cable. At 8 MHz, an SWR of 17 (which is not uncommon in broadband applications) will cause half of the transmitter's power to be lost in this coaxial cable.

Parallel wire line causes significantly less loss. The "matched" loss at 2 MHz is only 0.01 dB, rising to 0.05 dB at 8 MHz. This is a factor of ten less than that for coaxial cable. Further, the loss does not increase as rapidly with increasing SWR (see Figure 2). A 100-foot length of parallel wire line will not produce 3 dB of loss at 2 MHz until the SWR rises well above 500. At 8 MHz 3 dB is reached at an SWR of about 175.

Parallel wire line is commonly available with impedances ranging from 75 to 600 ohms. The most common impedances, approximately in order of popularity, are 450, 300, 600, and 75 ohms. Interestingly, if a 72 ohm antenna were fed with 100 feet of 72 ohm coaxial cable, the transmission line loss would be about 0.5 dB at 8 MHz. If this same antenna were fed with 450 ohm parallel wire line, the SWR would be about 6:1 but the line loss would be only 0.2 dB.

BALUNS

The term "balun" stands for "BALanced to UNbalanced" converter. They can be important in optimizing the radiation pattern and radiation efficiency of balanced (Hertz-type) antennas (e.g., dipoles) when they are fed from radios with unbalanced output circuits or over unbalanced transmission lines such as coaxial cable. Military doctrine does not make much use of baluns. In fact, their use is not important when each side of a balanced antenna is installed in a similar manner (e.g., at the same height above ground and near similar objects) and when the length of a coaxial feed line is not close to a half wavelength (or multiple thereof) of the operating frequency. When either of these conditions are not met a balun should be used. [7, 8]

ANTENNAS

An antenna in any radio system provides the link between the radio equipment and the space through which signals are sent.

There are many antenna designs. Many are useful only in specific situations. Some provide equally good (or sometimes, "equally bad") performance over a wide range of applications. In all instances, they are the portals between us and those with which we wish to communicate. They must be selected with specific objectives in mind.

In evaluating an antenna design for a specific purpose, NVIS communication or any other, there are two primary concerns:

• How much of the transmitter's power actually makes it into the radiated signal?

• Does the radiated energy go where needed?

In simple language, we are concerned with the antenna's radiation pattern (and thus its coverage) and with how much signal is radiated (its efficiency) over the range of frequencies required by the mission.

There are some practical concerns as well. In tactical situations, it is important to be able to erect the antenna quickly and

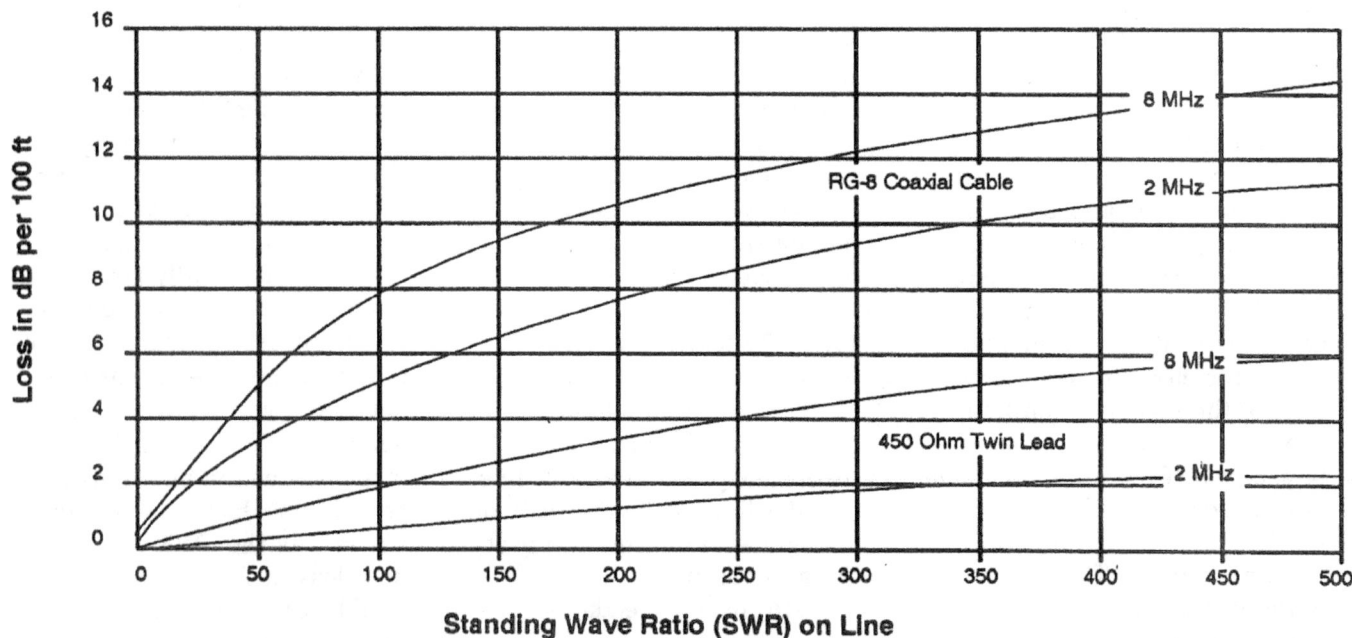

Figure 2. Transmission line loss.

easily. It must survive the weather and environment it encounters. There may be concern about its visibility to the enemy and the enemy's ability to intercept our signals.

Erecting an antenna quickly and easily is usually at odds with achieving optimal **mounting height** yet most antenna length above ground. As the height of an antenna, as measured in wavelengths, decreases its effective length becomes slightly longer and its feed point impedance becomes lower.

Antenna radiation pattern shape does not change much as an antenna is lowered below 0.1 wavelength although vertical gain decreases markedly. Fiedler [5] has provided an excellent description of the effect of antenna height on power gain (see Figure 3). Pattern shape does change significantly as an antenna rises above 0.25 wavelengths (see Figure 4 (all antenna patterns presented were prepared using [6])). A horizon-

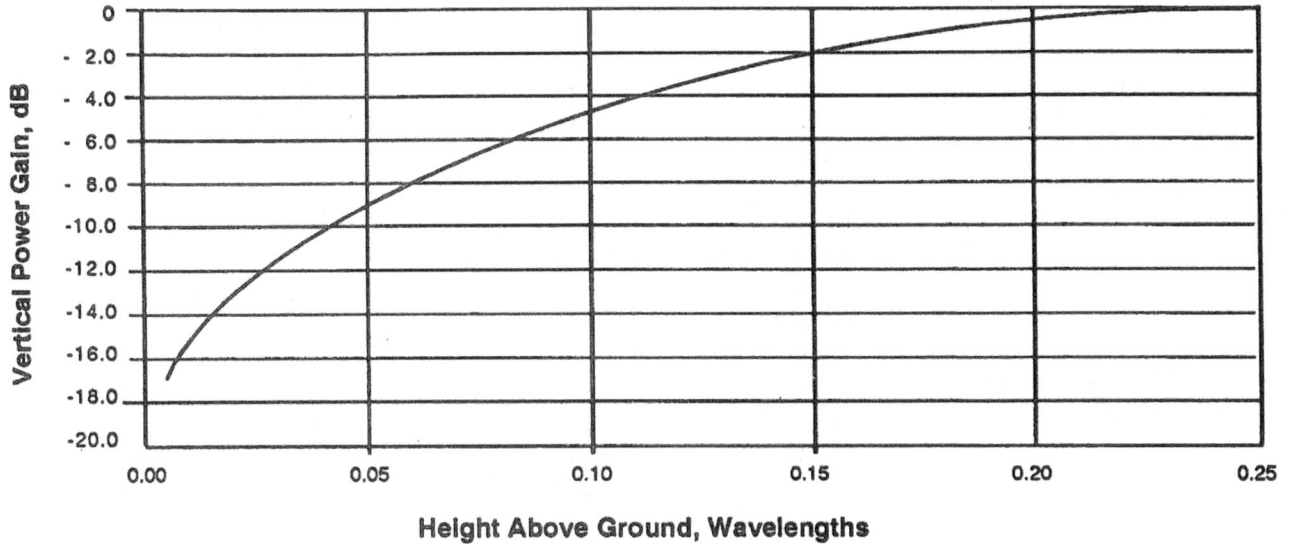

Figure 3. Antenna gain versus mounting height.

performance factors are strongly effected by it. Especially at the lower NVIS frequencies, it is difficult to attain mounting heights that optimize the NVIS effect (e.g., 0.1 to 0.25 wavelengths or 47 to 117 feet at 2 MHz).

All antenna dimensions are significant in terms of wavelengths. This is especially important when an antenna is to be operated over a range of frequencies. An antenna that is "low" at 2 MHz may be a long way in the air at 8 MHz. The antenna's impedance, gain, and pattern all depend on the mounting height in **wavelengths**.

The feed point impedance and antenna element lengths calculated with simple formulas are accurate only for antennas mounted at least a half wave-

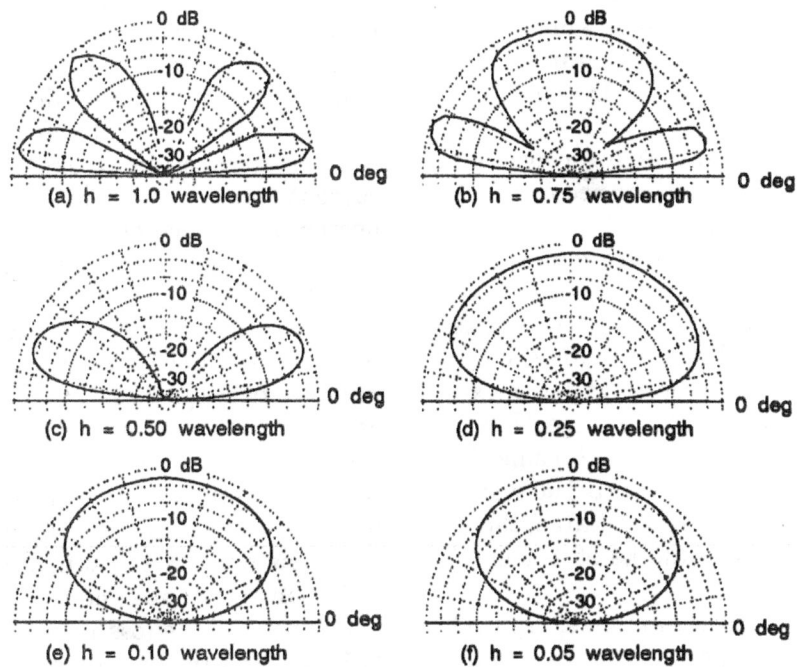

Figure 4. Vertical radiation patterns for a half-wave horizontal dipole at the mounting heights indicated. Note that NVIS effect is pronounced at h = 0.25 wavelength and remains so as mounting height is decreased.

tal dipole with excellent omni-directional NVIS characteristics when mounted at 0.25 wavelengths becomes directional and has no NVIS capability when mounted at a half wavelength.

These practical concerns are important because most performance objectives are optimized at the expense of physical size, complexity, and installation difficulty. Optimization exacts a particularly heavy penalty as frequency is lowered. Design innovation for operation at medium frequency largely amounts to finding ways to get performance without physical size and weight.

Let's look at the three central concepts in NVIS antenna design: efficiency, operation over the range of frequencies required by the mission (bandwidth), and radiation pattern (coverage).

Efficiency

The "efficiency" of an antenna describes how good it is at converting radio transmitter power into electromagnetic radiation. It does not address whether the radiation produced is useful for a specific (e.g., NVIS) purpose.

Each radio transmitter has some amount of power available at the antenna connector. Communication depends on how much of that power makes it into the antenna's radiation field. Numerically, efficiency is that which is useful divided by the total amount available, expressed as a percentage. For example, if it was determined that 95 watts from a 100 watt radio was being radiated, the efficiency would be 95/100 or 95 percent.

With most antennas, especially when operated over a broad range of frequencies, transmission line losses are larger than those in the antenna itself.

Antenna losses— When an antenna (such as a half-wave dipole) is described as having a feed point impedance of 72 ohms, what does that mean? If the antenna is resonant, it means that it exhibits a load at the feed point of 72 resistive ohms. There are two components of this resistance: radiation resistance and loss resistance.

• **Radiation resistance.** Some of the power sent into the antenna is radiated—it becomes the signal. Radiation resistance is a way of describing the antenna's success in converting the transmitter's output into radiated energy.

• **Loss resistance.** In the previous example, only 95 watts of the original 100 made it into space. Where did the other 5 watts go? One source of loss is the actual resistance of the antenna components (e.g., the wire and connections). We know that all wire has resistance and that resistance causes losses for radio frequency antenna currents. Some of the energy from the transmitter is wasted because the flow of the radio frequency current through wire and connection resistance produces heat instead of useful electromagnetic signal.

For a full-size half-wave dipole the loss resistance is a small number, perhaps less than an ohm, while the radiation resistance is quite large—72 ohms or so. The antenna's efficiency can be found by dividing the radiation resistance by the sum of the radiation resistance and the loss resistance. If the loss resistance is 1 ohm and the radiation resistance is 72 ohms, then the efficiency would be 72/73 = 0.99 or 99 percent.

Factors that effect efficiency include—

• Introducing one or more loading resistors. These increase the loss resistance.

• Improving the performance of a "short" antenna by adding loading coils. These coils make the antenna electrically longer than its physical length. This improves the radio frequency current distribution and thus increases the radiation resistance, but the substantial amount of wire in the loading coils also introduces additional loss resistance.

• Decreasing the distance between an antenna and the ground, decreases radiation resistance.

Antenna bandwidth

In practical implementations, it is important that the antenna operate over an adequate range of frequencies, either without operator intervention, or with only simple modifications. This is particularly important in systems that rely on Automatic Link Establishment (ALE) which "sound" many radio frequencies to automatically select the best one for a particular communications objective.

The range of frequencies over which an antenna is useful defines its "bandwidth." What is meant by "useful" depends on the available equipment and the specific communication objectives.

Standing wave ratio (SWR) bandwidth is a measure of the range of frequencies over which an antenna is "matched" to its source of supply (e.g., its transmission line). It is defined as the range of frequencies over which the SWR is within some range.

At resonance, the feed point impedance of an antenna is resistive. If the operating frequency is raised above the resonant frequency, the feed point impedance gains a capacitive component—the antenna presents an impedance to the transmission line that electrically "looks like" a resistor and a capacitor in series. On the other hand, if the operating frequency is below resonance the feed point impedance becomes inductive. The frequency, in either direction, at which the reactance is equal to the resistance is called the "3-dB point." It is also the point at which the SWR at the feed point will be 2 to 1, consequently, it is also called the "2:1 SWR point." The difference between the frequencies associated with the 2:1 SWR points (above and below the resonant frequency) is called the 2:1 SWR bandwidth. Bandwidth can be defined in terms of any SWR—there is no magic about 2:1.

Bandwidth is conveniently expressed as a percentage of the resonant frequency. For a typical high frequency wire antenna, it is about 5 percent of the frequency at which the antenna is resonant. At a resonant frequency of 2 MHz this provides a bandwidth of about 100 kHz. At 8 MHz an equivalent antenna would have a bandwidth of about 400 kHz. This means that the antenna does not have to be adjusted to maintain the SWR at less than 2:1 when the frequency is changed within these ranges.

Factors that effect bandwidth include—

• The larger the diameter of the wire the wider the bandwidth. An antenna made from #20 AWG will have less bandwidth than one made from #10 AWG.

• The shorter the physical length of the antenna the lower the radiation resistance and hence the lower the bandwidth.

• The lower the antenna is mounted the lower the radiation resistance, hence the lower the bandwidth.

As operation diverges from resonant frequency, the SWR on the transmission line increases. The ATU is designed to take care of this—it adjusts the antenna and its feed line to resonance. This permits operation over a broad frequency range. The feed line SWR is mainly of concern because of the resulting feed line losses. This is particularly true when coaxial cables are used.

Radiation pattern bandwidth has to do with the range of frequencies over which the antenna's radiation pattern is suitable for specific communication objectives. An antenna with excellent NVIS effect at 2 MHz may have no NVIS effect at 8 MHz. This leads us to the subject of "coverage."

Coverage

The usual NVIS propagation objective is to enable all stations within an area with a radius of 250 km or so to communicate with each other. This is fairly easy to do when an antenna can be tailored for each operating frequency. When a single antenna is to be stretched over what can be two-octaves (2 to 8 MHz) it becomes important to consider the frequency dependent aspects of radiation patterns in some detail.

Tactical considerations favor the simplicity of erecting a single antenna that can be used at all required frequencies yet any single antenna has dramatically

different radiation characteristics over such a two octave range. As frequency is increased the length and the mounting height both become a larger number of wavelengths. A dipole with excellent NVIS characteristics at 2 MHz may turn into a low radiation angle directional end-fire array with no NVIS capability at 8 MHz. There may be a frequency above which there is insufficient NVIS effect.

What constitutes an "adequate" range of frequencies depends on solar activity, the time of the year, and the mission [2]. At any position in the eleven year solar activity cycle, and in any season of the year it is usually essential that operating frequency be changed between night and day. This necessity is particularly acute during the shorter days of winter. In such circumstances, night operation will usually involve frequencies below 4 MHz while daytime operation involves frequencies above. [1, 2]

During the low portions of the solar activity cycle, night operation will frequently be in the vicinity of 2 MHz. Day operation will generally be below 8 MHz (in fact, will usually be below 6 MHz). An antenna that is a half wave at 2 MHz is 2 wavelengths long at 8 MHz. An antenna mounted between 0.1 and 0.25 wavelengths above ground (as suggested by [1]) at 4 MHz would be 0.05 to 0.125 wavelengths above ground at 2 MHz and 0.2 to 0.5 wavelengths above ground at 8 MHz. The radiation characteristics of a specific antenna installation will change substantially over such a range of frequencies.

Figure 4 shows the vertical radiation patterns for a resonant dipole antenna at various mounting heights. Figures 5 and 6 show the azimuth radiation pattern and the vertical patterns in two directions for a simple "Inverted L" antenna designed for 4 MHz. In Figure 5, the antenna is operated at 2 MHz while in Figure 6 it is operated at 8 MHz. Note that what was a good NVIS antenna at 2 and 4 MHz becomes hopeless for NVIS purposes at 8 MHz.

Antenna Radiation Patterns and Beamwidth— Evaluation of antennas for NVIS applications over a range of frequencies requires consideration of both the azimuth and elevation patterns.

The direction (azimuth) at which an antenna radiates is particularly important when a specific antenna is operated over a large frequency range. An antenna that provides a near circular radius of coverage at a low frequency (such as 2 MHz) might become directional with little or no NVIS effect at 8 MHz. See Figures 5 and 6.

The takeoff angle required to span a specific distance depends on the height of the ionosphere which, in turn, depends on the time of day, the time of year, and the position in the sun spot cycle. Reference [2] discusses this in detail.

An antenna's coverage depends on when and how it is

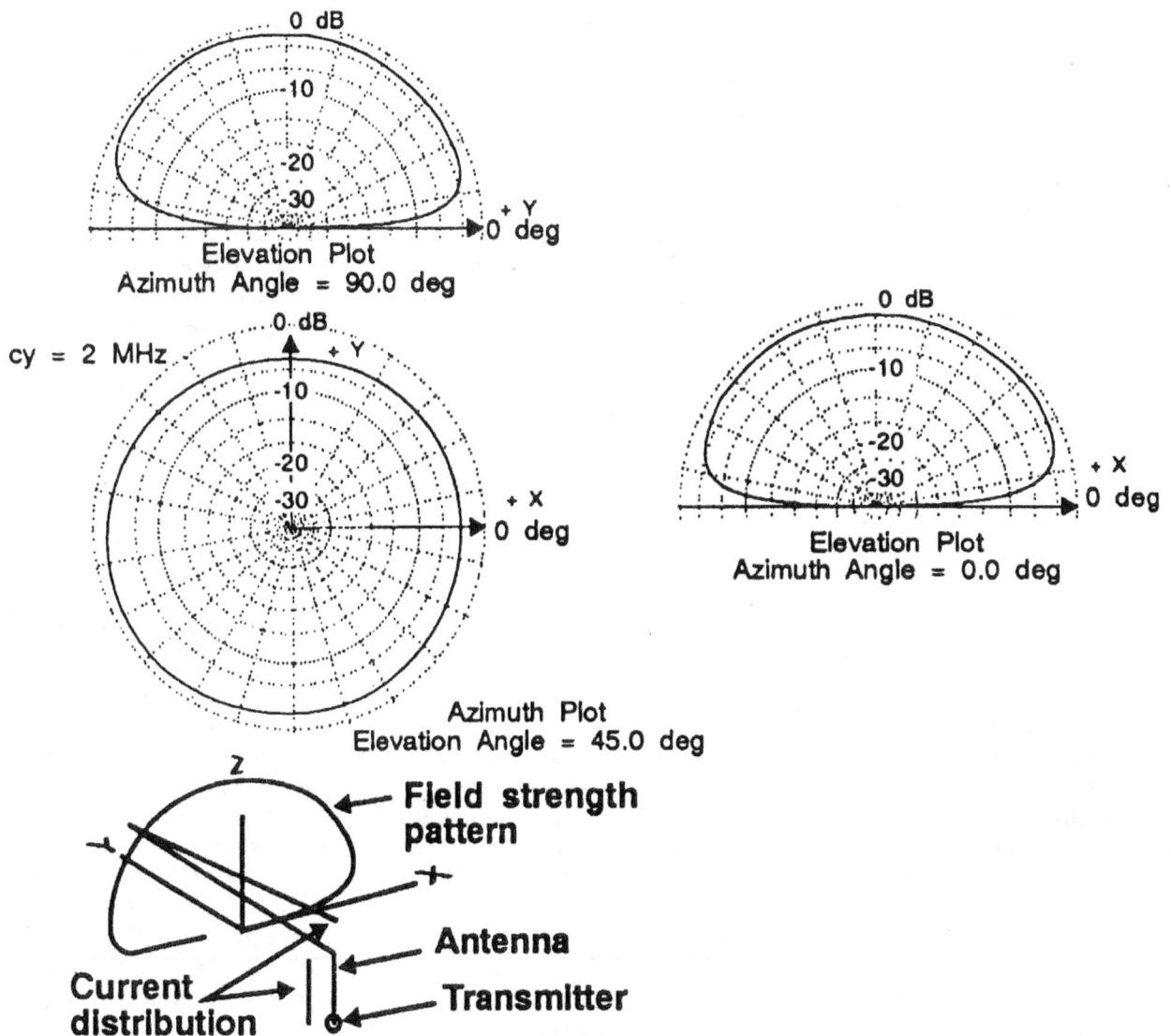

Figure 5. *4 MHz, half-wave Inverted L mounted 25 feet (0.1 wavelength at 4 MHz) above medium conductivity ground, operated at 2 MHz.*

used. While an antenna's radiation pattern can be evaluated what it means relative to a specific communication objective requires a broader understanding of NVIS propagation. [2]

Pattern plots are a useful tool in assessing coverage but it is convenient to have a "shorthand" way of describing these patterns. This can be accomplished by identifying three parameters.

First, there is the line (the *axis of directivity*) that shows where the largest portion of the radiation field is going. For NVIS work the preferred direction is straight up.

Second, on either side of this axis there is less radiation intensity. The points at which the radiation is half that along the axis (the point at which intensity is down by 3 dB) define the *beamwidth*. The beamwidth provides an indication of the area over which the antenna can provide "even coverage."

Third, it is useful to have some indication of the limit of coverage. While there are many criteria that could be used it is convenient to define the limit of coverage as the angles at which the antenna's pattern drops below one tenth that along its axis. This occurs at the -10 dB points on any pattern plot. This parameter, sometimes referred to as the *beam limit*, provides some measure of how easy our signals might be to intercept.

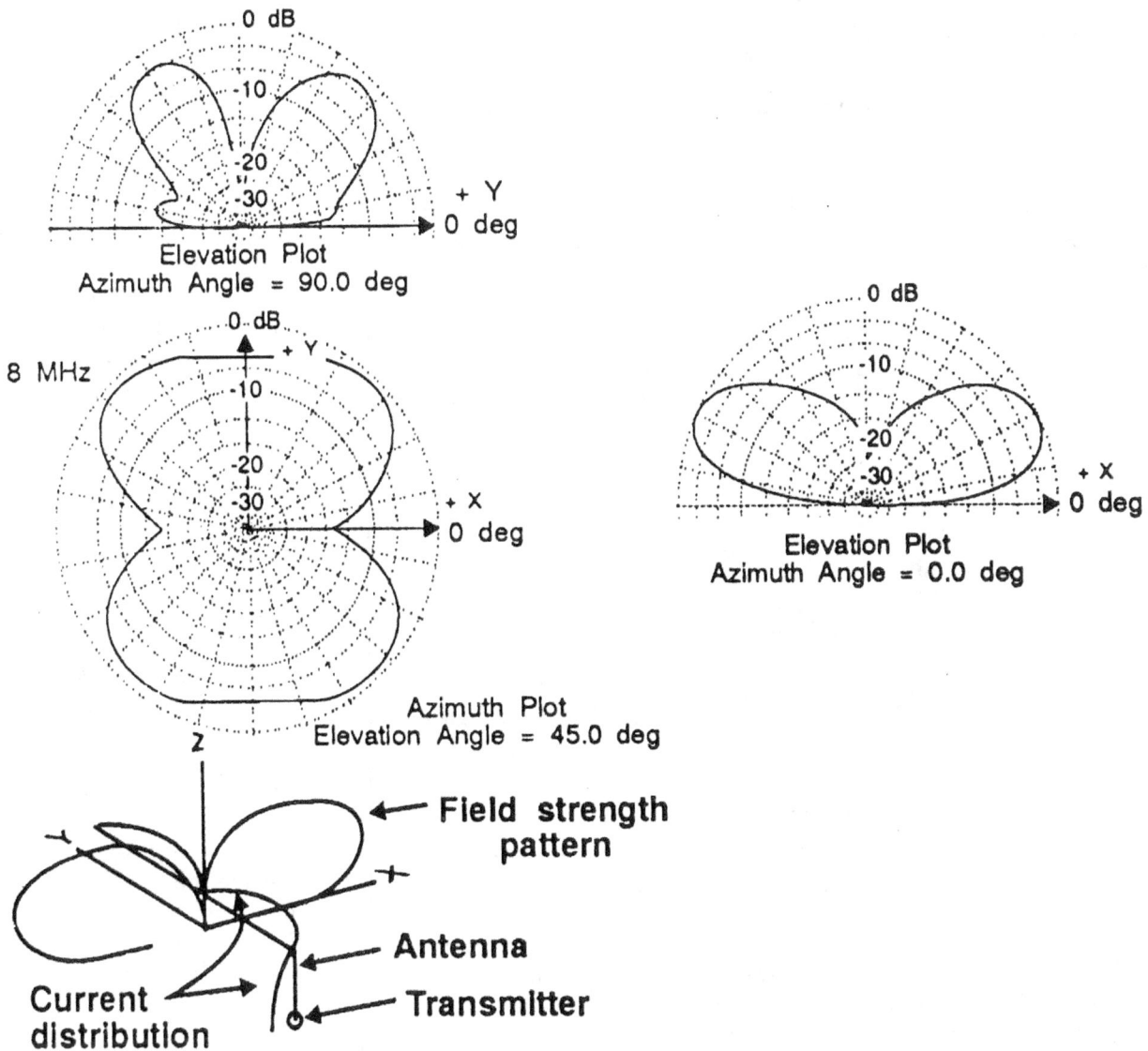

Figure 6. 4 MHz, half-wave Inverted L mounted 25 feet (0.1 wavelength at 4 MHz) above medium ground, operated at 8 MHz.

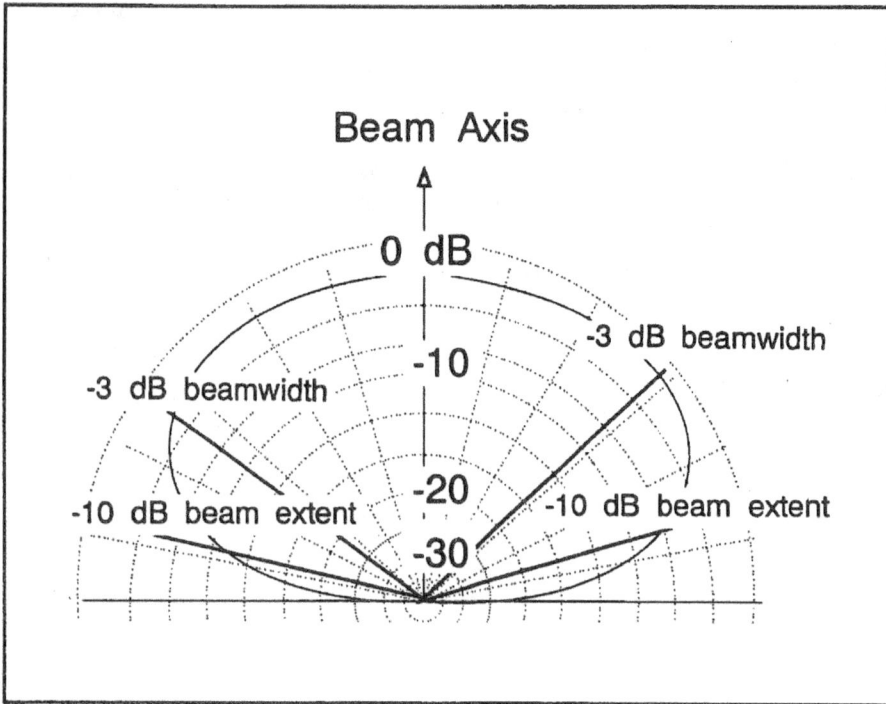

Figure 7. Pattern description parameters for a typical NVIS antenna.

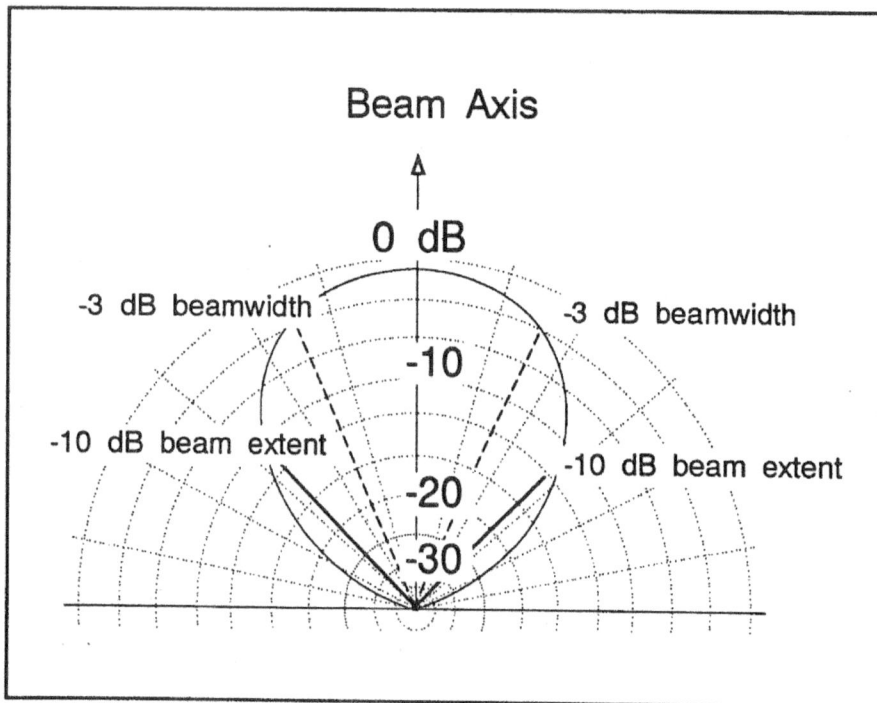

Figure 8. Vertical pattern of a three-element array based on the Shirley Dipole. Note the beamwidth and beam extent are much narrower than those for the single element half-wave dipole of Figure 7.

There is also a performance advantage in having restricted beamwidth. Noise arriving at the antenna due to interfering signals and atmospheric effects is attenuated in the same manner as is the transmitted signal. A good NVIS antenna inherently attenuates noise arriving at low angles.

For tactical NVIS work we would like to have a 3 dB beamwidth that is adequate to cover the area of interest and we would like the pattern to fall off very quickly (to -10 dB or better) as the angle increases beyond the 3 dB beamwidth.

Figure 7 shows the vertical radiation pattern of a half-wave dipole mounted about 0.1 to 0.25 wavelength above the earth. The 3 dB and 10 dB points are marked. Note there is very significant radiation well below 45 degrees. While this may be suitable for many applications there are those in which it is attractive to have much tighter patterns.

Figure 8 shows the vertical radiation pattern in a plane perpendicular to the array's axis for a three-element array based on the Shirley Dipole Array. Note the much sharper beamwidth.

An Interesting Example—

Consider a real mission of the California State Guard and the California Governor's Office of Emergency Services. Some of the sites that could be involved in emergency communications include the Capitol of California in Sacramento; Stockton; Reno, Nevada; Fresno; and Long Beach. See Figure 9. These paths are described in Table 1.

Table 1. Path geometry relative to Sacramento.

Location	Distance, km	Azimuth, Degrees	Takeoff, Degrees
Stockton, California	72	162.0	82
Reno, Nevada	185	53.2	71
Fresno, California	248	140.4	65
Long Beach, California	602	149.1	40

The takeoff angles shown are for "worst case" winter conditions [2].

Suppose the 4 MHz Inverted L antenna of Figure 5 is erected so its "Y-axis" is aligned with a straight line between Sacramento and Long Beach. At 4 MHz, this antenna provides mission-compatible coverage.

Now, suppose conditions require operation at 8 MHz. This same antenna becomes directional (see Figure 6) and an operator in Sacramento would find communication with Stockton and Reno impossible, Long Beach difficult, but Fresno quite good.

If the same antenna (still operating at 8 MHz) were rotated 90 degrees, it would be found that it was still impossible to communicate with Stockton, difficult with Fresno, but easy with Reno and very easy with Long Beach.

Remember, these results were achieved with what was designed as a good NVIS antenna with a near-circular coverage area at 4 MHz. Clearly, when operation over a range of frequencies is required, changes in antenna radiation pattern must be considered.

Figure 9. Locations involved in the California example (see Table 1 and text).

References

[1] US Army Field Manual, FM 24-18 *Field Radio Techniques*; Appendix N; December 1984.

[2] Farmer, Edward J., "NVIS Propagation at Low Solar Flux Indices," *Army Communicator Magazine*, Spring 1994, Fort Gordon, GA.

[3] Caron, Wilfred N., *Antenna Impedance Matching*, pp. 4-1 through 4-11, American Radio Relay League, Newington, CT, 1989.

[4] Maxwell, Walter M., *Reflections*; Chapter 4, "A View Into the Conjugate Mirror," American Radio Relay League, Newington, CT, 1990.

[5] Fiedler, David M., "HF Radio Communications and High Angle Antenna Techniques," *Army Communicator Magazine*, Summer 1991, Fort Gordon, GA.

[6] Lewallen, Roy, W7EL, *ELNEC Computer Program and User Manual*, Version 3.07, PO Box 6658, Beaverton, OR 97007, 1993.

[7] Lewallen, Roy, W7EL, "Baluns: What They Do and How They Do It," *Antenna Compendium*, Volume 1, The American Radio Relay League, Newington, CT, 1985.

[8] Roehm, Albert A., W2OBJ, "Some Additional Aspects of the Balun Problem," *Antenna Compendium*, Volume 2, The American Radio Relay League, Newington, CT, 1986.

Mr. (MAJ) Edward J. Farmer (CA SMR), a professional engineer, is president of EFA Technologies, Inc. The former signal soldier has a BS in electrical engineering and an MS in physics, both from California State. He has published over forty articles and two books. He also holds two US patents.

PART TWO

HOW TO DO IT

Military experience and the efforts of some emergency response organizations provide the best information concerning how to make NVIS work. A variety of situations are covered in the articles included in this section.

5. The HF NVIS Radio Path — An Engineering and Operational Challenge by Dr. Alan S. Christinsin, P.E.; from **Tactical HF Radio Command and Control — An Anthology, Volume I — Low Power, Short Path HF Communications**; Published by ASC & Associates Ltd., 1201 Dawn Dr. Belleville, IL 62220; 1993.

6. Near Vertical Incidence Skywave (NVIS) Propagation, The Soviet Approach by LTC David M. Fiedler, **Army Communicator** magazine, Spring 1987.

7. Mobile NVIS: The New Jersey Army National Guard Approach by LTC David M. Fiedler, **Army Communicator** magazine, Fall 1987.

8. Optimizing Low Power High Frequency Radio Performance for Tactical Operations by LTC David M. Fiedler, **Army Communicator** magazine, Spring 1989.

9. HF Radio Communications and High Angle Antenna Techniques by LTC David M. Fiedler, **Army Communicator** magazine, Summer 1991.

10. How to Survive Long Range and Special Operations by LTC David M. Fiedler, Army Communicator magazine, Summer 1993.

11. Antenna Performance for NVIS Communications by MAJ E. J. Farmer, **Army Communicator** magazine, Summer 1996.

12. U.S. Army FM 24-18: **Tactical Single-Channel Radio Communications Techniques**, Chap. 3 & Appendix M: Near-Vertical Incidence Skywave Propagation Concept, September 1987.

13. NVIS Refresher by Stanly Harter KH6GBX, *Worldradio* magazine, January 1995.

5: THE HF NVIS RADIO PATH

An Engineering and Operational Challenge

by Dr. Alan S. Christinsin

Since World War II, military communicators using high frequency (HF) radio over short skywave path lengths, also known as near vertical incidence skywave (NVIS), have had engineering problems. Congestion of the HF portion of the radio spectrum, combined with the broader bandwidths of today's independent sideband multichannel operations, make a sometimes difficult situation even worse. The purpose of this paper is to acquaint communicators with the problems associated with the HF NVIS path. A short HF skywave path is defined by most as being a distance from 0 to about 300 miles (500 km) in length.

There are several major problems associated with short-haul HF systems. Generally speaking, the shorter the path, the lower the frequencies used. The lower the frequencies employed, the greater the interference potential due to spectrum congestion, the increased degradation due to atmospheric noise (i.e., the worse the signal-to-noise ratio – SNR), and the lower the antenna gain and directivity. High noise, in itself, can prevent successful communication, especially at certain times of the year at specific locations in the world. This is particularly true in the late afternoon and evening. This is when atmospheric noise increases due to propagation of the static from thunderstorms located in southern latitudes. Increase of noise is caused by the absence or diminishing effects of the changing "D' layer (the dominant ionospheric absorption layer). Summer evenings are particularly noisy from about 1600-2200 local time due to the higher probability of local thunderstorms as well as the addition of propagated atmospheric noise from the equatorial areas. Refer to Figure 1.

Most HF equipment used on NVIS paths is fairly low power (usually less than a kW...often only 20 Watts). This isn't much power when competing with high noise and interference levels present on short radio paths. However, most HF operation at these short paths is by tactical users using vehicular or manpack HF radios. While it is always advantageous to have high power, in a tactical situation, transmitter power is usually limited by mobility constraints, fuel availability, and other considerations. As a result, few tactical HF systems use high power HF transmitters. The most successful communications over the HF short haul are usually obtained by employing communications

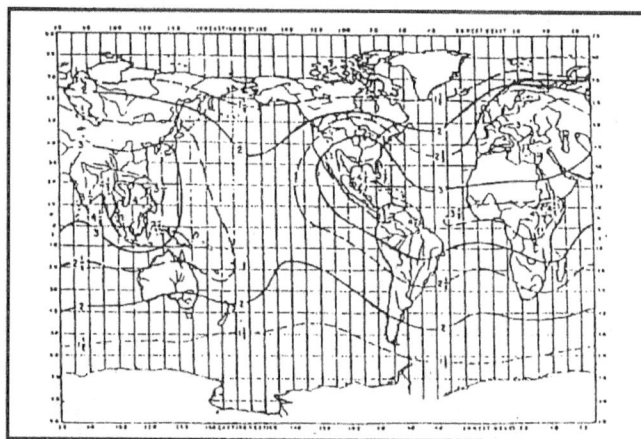

Figure 1. Atmospheric Noise Map-of-the-world (Summer in Northern Hemisphere)

modes requiring the lowest signal-to-noise ratios. If military units were regularly using International Morse Code (IMC— also known as CTI) on a regular basis, it would be very successful from an SNR standpoint because of its narrow (100 Hz) bandwidth and inherent excellent signal-to-noise ratio — *albeit low communications throughput.* For many years, the most common HF emission has been a single sideband voice (3 kHz) signal with suppressed carrier. This was primarily because voice operation has been very popular for military use for many years. The SSB receiver passband is about 3 kHz, therefore, only 3 kHz of noise is present. Since the power of a single voice channel is concentrated within 3 kHz, it competes for the dominance of the passband (i.e., the SNR) and, if reasonable signal levels are received, good SNRs are usually achieved. In recent years, however, new data transmission capabilities have been developed that operate at signal-to-noise levels as low or lower than the SNR for IMC with throughput up to 2400 baud at times. The greater the noise, the greater signal power density in the receiver's passband is needed to overcome the noise and produce a usable output. The new "smart" data modems can improve data operation dramatically. On the other hand, voice requirements are still difficult to satisfy at this point in time, but technology is improving in this area. The relatively high signal-to-noise ratio requirements of voice operation are difficult to achieve on short skywave paths. However, with the advent of more effective RF speech processors and digital signal processing devices to reduce the effects of noise and interference there have been significant improvements in voice operation in the last few years. While one cannot achieve SNRs for voice as low as with data, there will always be a requirement for both voice and data. There is more use of data now than heretofore because of its improved reliability and ease of encryption.

Since tactical transmitters usually employ fairly low power (usually 20 to 400 Watts), the best opportunity for improving the signal between two stations operating over a NVIS path is through the selection and use of properly engineered antennas at each station. Most communicators think of the Rhombic, Vee, Sloping Vee, Log Periodic, Yagi or Sterba curtain antenna when considering use of an *effective*, high gain, HF antenna. The aforementioned antennas are excellent, but although they have high gain, unfortunately, most are not designed for use over the NVIS path. At these short paths, lower frequencies and higher takeoff angles are required. Also, high gain antennas are very large and are not easily transported. The main problem is, however, that most of the aforementioned antennas produce low radiation angles (below 20 degrees) at higher HF frequencies. Although desirable for the long haul, they are very ineffective when used on a short path. The following chart (Figure 2) shows distances versus optimum antenna takeoff angles that should be considered when selecting an effective antenna for various path lengths.

Based on the foregoing chart, one can see that for short distances, takeoff angles are quite high...usually from 45 to 90 degrees above the horizon.

The shorter the HF skywave path, for all practical purposes, the lower the frequency of operation at a given time of day. To carry that one step further — the shorter the path, the lower the frequency and usually the longer and/or larger the antenna needed. The tactical communicator must reason a step beyond that. The longer the antenna for the short path, usually the higher above ground it should be installed for optimized performance.

For all practical purposes, there are no tactical "high gain" antennas for the short skywave path. A horizontal dipole a quarter wave above ground is going to produce about as much gain at 90 degrees above the horizon as is going to be obtained for *tactical* NVIS use. (The doublet antenna near a quarter-wave above ground produces a gain of about 5.6 dBi.) There are a very few large high angle tactical antennas with fair gain available commercially, but most have gain about only slightly greater than the doublet; however, unlike the doublet, they are broadband structures. Some of these antennas look like a regular log periodic struc-

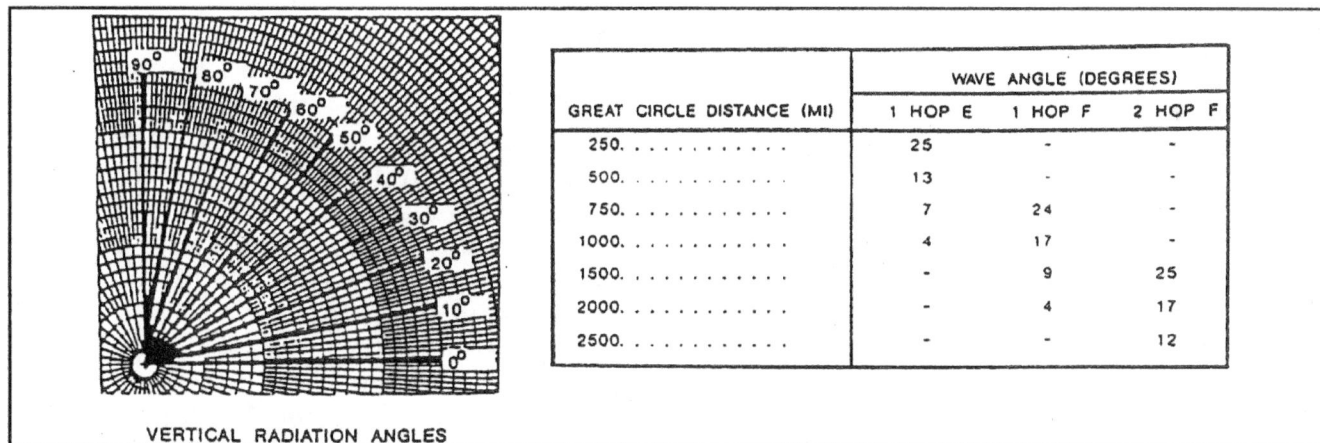

GREAT CIRCLE DISTANCE (MI)	WAVE ANGLE (DEGREES)		
	1 HOP E	1 HOP F	2 HOP F
250.	25	-	-
500.	13	-	-
750.	7	24	-
1000.	4	17	-
1500.	-	9	25
2000.	-	4	17
2500.	-	-	12

VERTICAL RADIATION ANGLES

Figure 2. Typical Required Vertical Radiation (Takeoff) Angles for Various Distances in the Northern Temperate Zone

ture only aiming straight up or siting on its short end aimed straight at the ground. They focus their radiation at the ground and reflect back toward the sky. An excellent log periodic antenna of this design is made by Antenna Products Corporation. It is a vertically aligned horizontal log periodic which is about 70 feet tall and about 120 feet long, called the LPH-15. This antenna covers the band 4-30 MHz with 5 dBi gain or more at all frequencies over the band. Although called a tactical antenna, it is large and almost impractical for tactical use. In fact, it is more properly considered moveable and takes about a day, maybe two, to install. However, it is an excellent NVIS antenna. Refer to Figure 3.

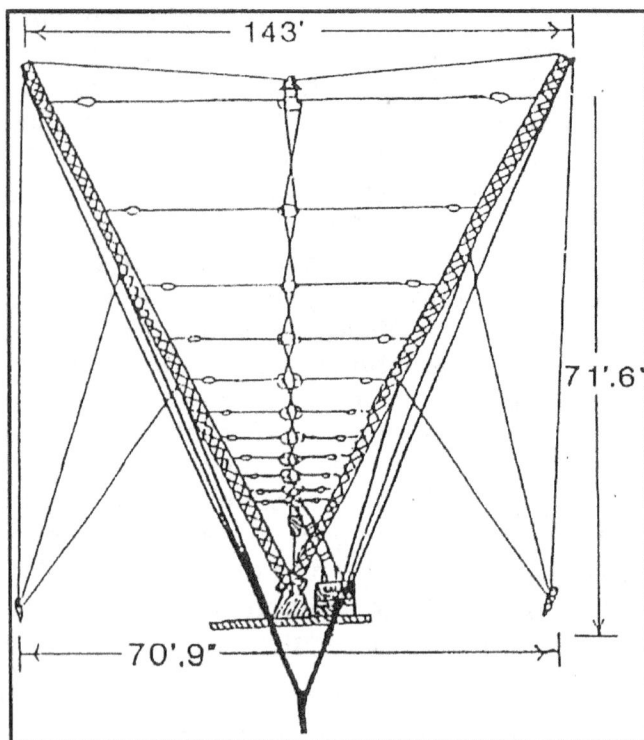

Figure 3. The LPH-15 Tactical Vertically Aligned Horizontal Log Periodic High Angle Antenna for Short to Medium Distances

The Doublet (resonant dipole) antenna is simple, easy to erect, and provides a fair amount of gain at the desired radiation angles. This is about 5 dB over an isotropic antenna (dBi) at heights under a quarter-wave above ground. Unfortunately, the doublet is a single frequency antenna (plus or minus 1 or 2% bandwidth of its design frequency), and it must be reconstructed/modified for each frequency change beyond one or two percent. The length of a dou-

blet is another problem. A doublet for 2 MHz would be about 234 feet long. This is a large antenna for tactical applications. In today's world, it is rare that single frequency antennas (like doublets) are employed. For one thing, with ALE, frequency hopping (and easily tuned radios) are impractical to use with single frequency antennas — or downright impossible. Also, the reader is reminded that the classic figure eight (bi-directional) azimuth pattern of a dipole applies only applies in free space and, to some degree, at height multiples of a half-wave above earth. At doublet heights close to earth (a quarter-wave or less), a near-omnidirectional pattern is formed (especially at the higher radiation angles). The greatest problem of the classic resonant doublet (half-wave horizontal dipole) is, of course, its narrow bandwidth as mentioned previously. Refer to Figure 4.

If only day, night and transition frequencies are used (three frequencies), the "Spider Web" (multiple doublet) antenna is a reasonably good compromise antenna for the short skywave path. However, it is limited to just the three specific frequencies. Its gain, patterns, and general performance are close to that of three classic doublets. It is fed with a single coaxial feed line. The "Spider Web" normally employs only one set of masts (a set normally consists of three masts — or poles). Because the height of all three frequencies is fixed, at the one height there is some pattern compromise. Refer to Figure 5.

Another tactical antenna, the "Inverted L," has good gain at high angles (only slightly less than a dipole) and is frequency agile. This antenna, unlike the resonant dipole (doublet) is fed through an antenna coupler into a single wire feed line, making the antenna very frequency flexible. The antenna coupler is a component part of the radio package in most instances. The azimuth pattern and takeoff angles are somewhat similar to a doublet. Beyond about four times the lowest design frequency, however, the gain falls off sharply since the antenna starts to function like a long wire at one wavelength or more (i.e., maximum energy starts to radiate from near the ends rather than broadside to the horizontal wire). At the higher frequencies, the vertical takeoff angles become lower because the antenna's height remains constant in feet, but changes in wavelength.

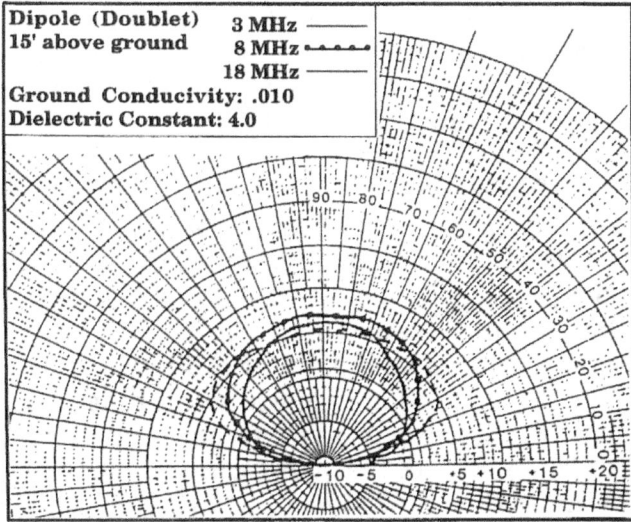

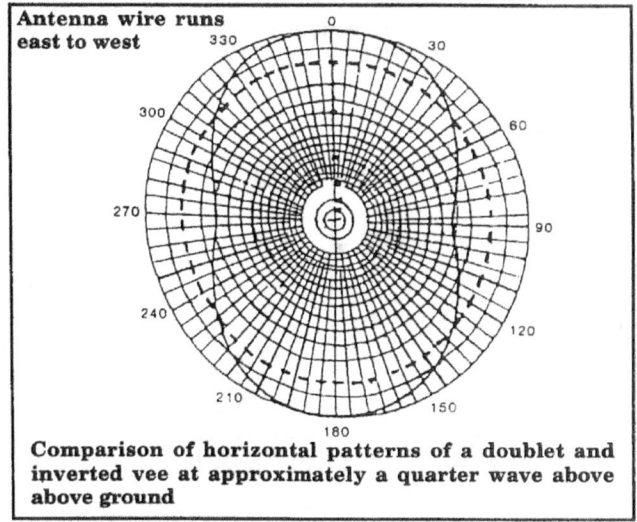

Figure 4A (left) and 4B (right). The Azimuthal Radiation Pattern of a Doublet antenna a quarter-wavelength or less above ground.[2]

See Figure 6.

A variation of the doublet is the "Inverted Vee" antenna. The "Inverted Vee" is essentially a dipole supported by a single mast or pole and its legs 'drooped" towards ground. *It is a single frequency resonant dipole*, but, like the doublet, can be broken into segments with jumpers and insulators so that up to three specific frequencies can be used if the jumpers are changed for each of the three frequencies. This is a useful capability, but, again, the number of frequencies available is very limited and use of jumpers requires lowering the antenna in order to change them for each "QSY" (frequency change). Refer to Figure 7.

Figure 5a. Fort Monmuth design "Spider Web" antenna

Another NVIS antenna is the Rockwell-Collins orthogonal antenna often referred to as the "NVIS Antenna" — the 637K (the Collins model number). This antenna is now widely known by its AN nomenclatures, the AS-2259 or 2268. Actually, this antenna comes in a couple versions, hence, the different numbers. Both the AS-2259 and 2268 consist of two crossed sloping dipoles connected to each other at the feed point, each dipole is cut to a different frequency, and the antenna structure is tuned at the base of the mast. The mast is made of rigid coax and is tuned using an antenna coupler. (A component of the radio set.) It uses the same antenna base and the same antenna coupler as a whip. The AS-2259 and 2268 were specifically designed to operate over NVIS paths, but using hardware peculiar to a whip antenna. The antenna has reasonable cost, produces less gain than a resonant dipole, and is another pack to carry. However, it is a good NVIS antenna, it does function well, and has been used for many years. It consumes a fair amount of space—in a specific geometrical configuration. This antenna radiates its maximum vertical lobe directly above from about 90 degrees down to about 50 degrees — perfect for NVIS operation. It is designed to operate from 2 to 12 MHz. While it can be tuned to higher frequencies, the pattern becomes unpredictable ("flaky") and should not normally be used at these higher frequencies. Your attention is directed to Fig-

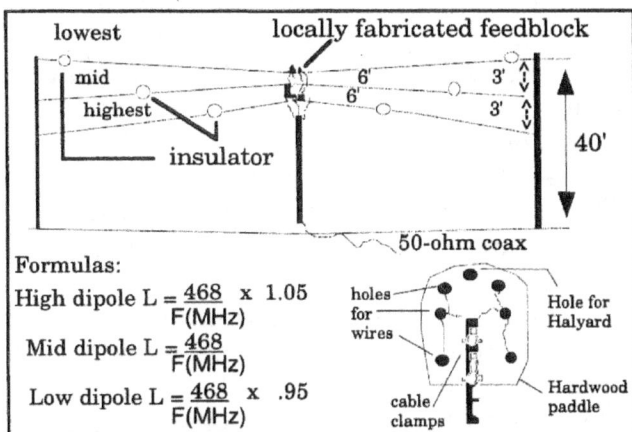

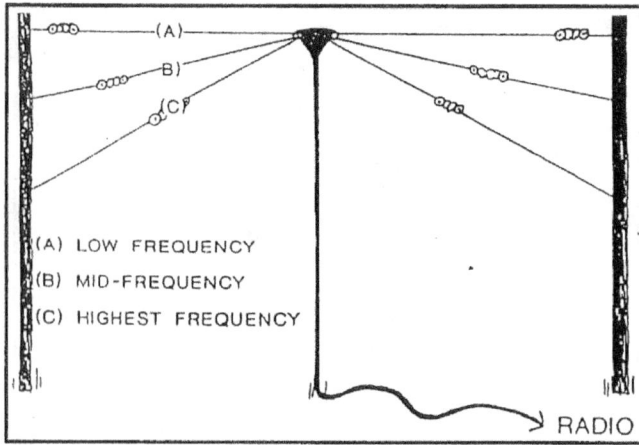

Figure 5b. Basic design "Spider Web"

(A) LOW FREQUENCY
(B) MID-FREQUENCY
(C) HIGHEST FREQUENCY

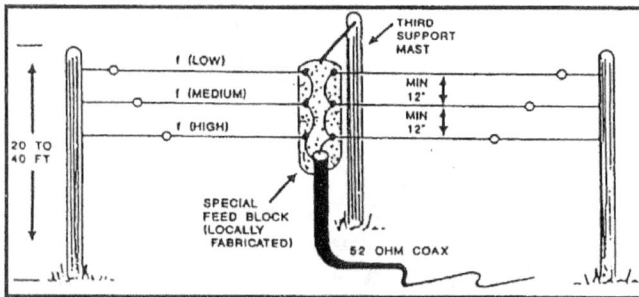

Figure 5c. Stanford Research Institute (SRI) design "Spider Web"

ure 8.

For semi-fixed or transportable use, the T2FD antenna (a broadband dipole) has seen considerable use in the last 8 years — especially in the Gulf War. The Barker and Williamson (B&W) broadband dipole, which is actually a compromise version of the classic "Terminated 2-Wire Folded Dipole (T2FD)," has performance almost as good as the classic T2FD. The T2FD was designed by the Naval Research Laboratory (NRL) back in about 1949 and has been used by the Army, Navy and Air Force as well as the National Security Agency (NSA). It is now made commercially by B&W in a modified configuration for a cost of around $200.00. It basically functions and has patterns like a resonant dipole (doublet), but it has very wide bandwidth. However, after about 4 or 5 times its lowest (design) frequency, the vertical pattern tends to lay down somewhat and the azimuthal pattern's maximum radiation moves from broadside to the plane of the wire to off the ends of the wire. The antenna is not especially efficient (40% or less compared to the efficiency of a resonant dipole which has over 70% efficiency at the height of about 25 feet above

ground). However, the T2FD has the advantage of being a high angle broadband antenna with the capacity of supporting automatic link establishment (ALE) and frequency hopping anti-jam (AJ) operations. The classic T2FD, and the B&W as well, are a bit bulky to deploy because of the separated wires and spacers making these antennas difficult to roll up and transport. For NVIS use, install the antenna less than a quarter-wave above ground (at its lowest design frequency). Refer to Figure 9.

What is the message after all that discussion? First, there is no perfect antenna for HF NVIS operation. Secondly, compromise is the name of the game. Third, what's good for one short haul situation, is not always good for another. The best advise is to size up your situation and determine the type antenna that will

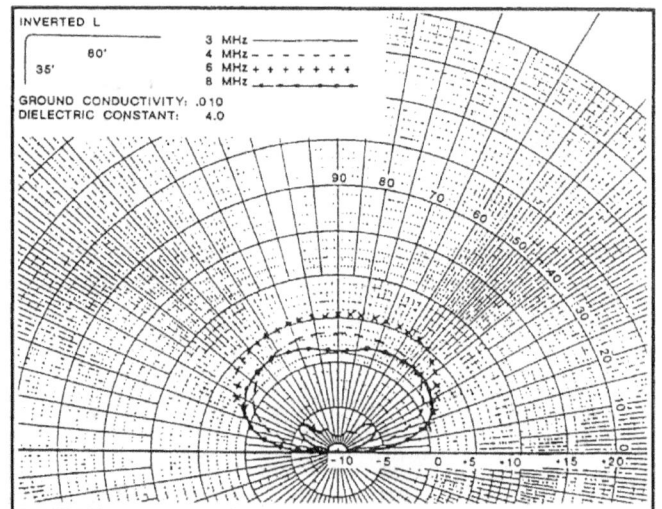

Figure 6. The Inverted L antenna

do the best job for your requirement — considering all the specific constraints. In this modern age, the trend is away from resonant dipoles (doublets) and more to frequency agile antennas such as the Inverted L (tuned antenna) and broadband antennas such as the T2FD and others because of their usefulness for automatic link establishment (ALE) and frequency hopping.

Over the short haul, from a propagation standpoint, not many frequencies are required — usually three (six for a short distance duplex system). This usually consists of a "day" frequency, a "night" frequency, and a "transition frequency for a total of three for a simplex system. Six are required for duplex operation.

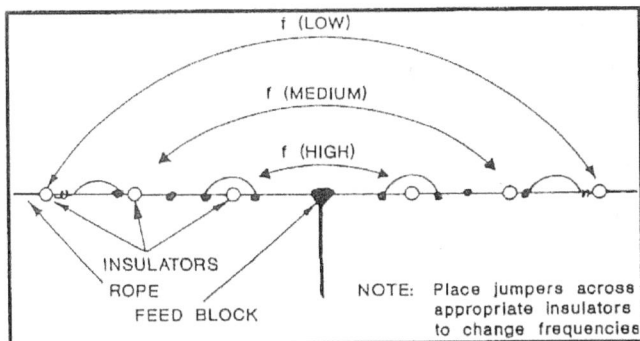

Figure 7. The Jumpered Multi-frequency Doublet

The usage of these frequencies on an hour-to-hour basis in a multipoint simplex net is determined by the net control station (NCS) if a manual system (multipoint net). In the low end of the solar cycle, short path frequencies are generally confined to the 2- to 8 MHz band because of the low solar activity while during SSN maximum it might be 2-12 MHz. A typical (simplex) frequency complement for a 100 mile path at SSN minimum, winter might be 7 MHz (day), 4 MHz (transition) and 2 MHz (night). There are several problems associated with propagation and frequency usage at SSN minimum. One is that during SSN minimum, the lower portion of the HF spectrum becomes very congested — with all users of the spectrum moving lower and lower in the HF spectrum as the solar activity decreases. The result is that users increasingly operate on lower frequencies. This increases occupancy in the lower portion of the HF bands, causing increased interference, especially at night.

Those users who have lower HF frequencies already assigned will start using them regularly — even when not actually needed, as

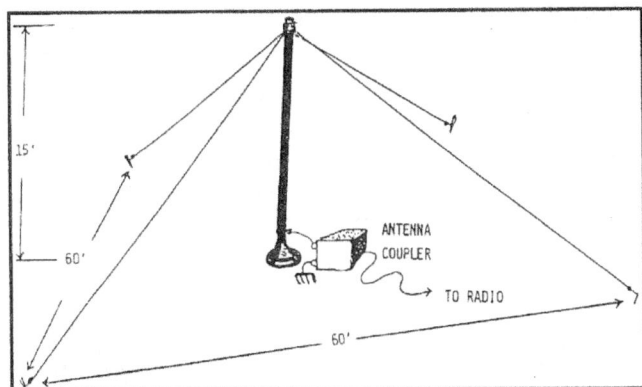

Figure 8. The AS-2259/2268 (also known as the Collins 637K)

soon as they realize that SSN minimum is drawing near. The reason is often to discourage other stations from encroachment. Because of heavy spectrum occupancy, it is difficult to obtain additional low HF frequency assignments when requested, and if they are actually assigned, usually they are authorized on a non-interference basis (NIB) and "clobbered" with interference (probably along with those frequencies the station already has assigned). NVIS frequency support has historically been a problem, especially *during SSN minimum* where interference is a major consideration. Refer to Figure 10.

Interference has been discussed throughout this paper, so there isn't much more that can be added. The real variable in HF communications, over which we have little or no control, is interference from an outside source (other stations using the same or adjacent frequencies). Additionally, on-site interference

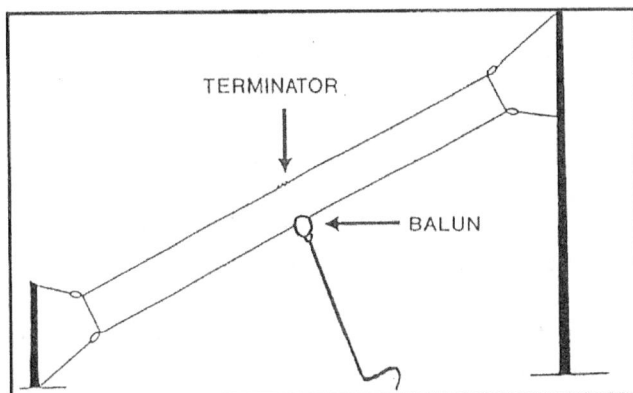

Figure 9. The T2FD (Terminated 2-Wire Folded Dipole) Antenna Configured as an Inclined Version

problems can be created due to poor siting mostly with respect to other closely collocated electromagnetic radiating and receiving equipment. There is usually no difficulty convincing tactical communicators to separate their "Sloping Vee" or other long haul antennas by 500, 1,000 feet, or more. However, much more serious coupling is possible between two doublets using lower HF frequencies (or worse yet, two whips operating over short-haul ground-wave paths). Yet, often such antennas are located almost next to each other. Antenna selection and placement exert a considerable influence upon use of assigned frequencies. Often the interference produced by collocated antennas

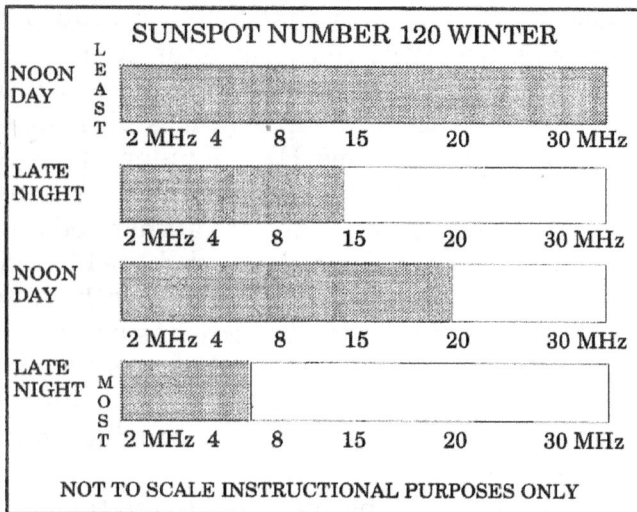

Figure 10. Congestion of the HF segment of the radio spectrum.

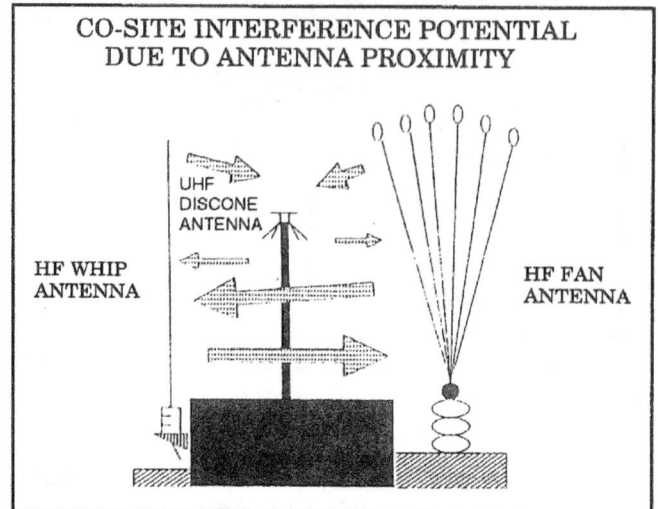

Figure 11. Electromagnetic co-site problems caused by Antenna Proximity

prevents the use of some assigned frequencies at certain times and this can degrade system reliability greatly. Refer to Figure 11.

Location of receive equipment near high tension power lines can cause pick up of considerable man-made noise from that source. The cumulative results of man-made and atmospheric noise compound the ambient noise problem, at times, making effective communications shaky at best. Sometimes reliable communications cannot be achieved for many possible reasons.

Often, no matter what is done, it may be impossible to make the short skywave/NVIS operation work reliably during some periods of time...especially on some nights. In some cases, one might elect not to employ NVIS operation because of the severe levels of noise and interference and instead operate over a long path to a more distant station and have the traffic relayed back to your station on the longer system. For example, instead of working from one part of Florida to another 100 miles away, it may make more sense to operate from Florida to New York State (a longer path) and then relay back to Florida. This would reduce atmospheric noise considerably, possibly reduce interference from spectrum congestion on the lower frequencies, reduce the size of antennas used, and probably improve overall system reliability. Because voice is usually impractical to relay reliably, only data systems would be candidate for relay. Refer to Figure 12.

Usually the operational situation prevents this approach and one may be committed to operation over a short skywave path even with all its problems. If so, one should ascertain how best the system can be engineered to squeeze as much reliability out of the system as possible. That is, select the best available hardware, pre-site the facilities (especially the antennas), configure the end stations properly, employ trained and knowledgeable operators, and, if possible, obtain frequency redundancy (more than a bare essential quantity of frequencies in the required bands) in order to "dodge" interference. Although this is easy to say, it is not always easy to do. The operator should do the very best he/she can to minimize system losses, and received interference and noise.

A path reliability study should be performed in advance of the deployment, or at least as soon as possible. This will provide a feel for the anticipated path reliability and will point out the times of the RADAY when communication might be difficult or impossible. While such a study does not provide absolute solutions, it does provide a good starting point by focusing attention on potential problems. With it, the knowledgeable HF engineer can come up with a reasonable assessment of the probable reliability over the path and optimize, or at least enhance, the chances of success through knowledgeable adjustments to equipment complements, frequency assignments, antennas, etc.

As a result of the predicted path reliability (PPR) study, at least the period(s) that your system may "go out" can be predicted. Operators should be briefed about the situation and should be provided methodology on how to "squeeze" as much "up-time" out of the system as possible (i.e., tips on QSYing). Determine if a lower grade of service can be used by the subscriber, i.e., can the subscriber stand a noisier voice or degraded data with a higher BER than specified for the channel. If so, the system reliability estimate can be improved. If not, then there may be problems. Brief the system users that if this is the case, then it is highly probable that the station may have a traffic backlog between certain hours. This could allow internal personnel adjustments to be anticipated in advance because, when the system finally comes back in, things get hectic both in the communications center and for the administrative processing distribution people as well. At the same time, initiate action to upgrade the system to make it perform at a higher degree of reliability.

Even though studies might indicate that one may be reasonably certain that the HF NVIS radio system of interest could "drop out" at a certain time, there is no reason to give up. There are many operating techniques that can be employed to prolong the usefulness of HF radio communications...one being: going to a mode of operation requiring a lower SNR such as CW (S/N density ratio of about 34 versus 48

for order wire quality voice. Also, messages can be sent twice and other techniques could be used. Operators should be trained in these techniques.

The pre-deployment PPR (predicted path reliability) analysis will give the engineer a feel for the expected performance of the short haul HF system. Take all necessary steps to optimize the terminal engineering during the siting phase. Install equipment and antennas properly. Be certain everything is functioning correctly. Also, develop and post a propagation chart/"QSY" (frequency change) guide for the operator on duty to use (if manual operations are employed). For ALE controlled operations, a realistic frequency complement should be programmed into the ALE controller. Supervisory personnel might consider dropping by the station for a visit during the expected bad periods and offer encouragement to the operators and see the results first hand. Be certain that in a multipoint net that the NCS operator(s) on shift during times of projected low reliability be the most qualified and best trained.

Provided operators keep the necessary records for operations, post action analysis can be accomplished to identify problems of the foregoing nature. Post action analysis of the HF system is particularly important if similar operations are expected to improve in the future. A proper analysis can verify the validity of the pre-deployment path engineering predictions and assist in problem solving.

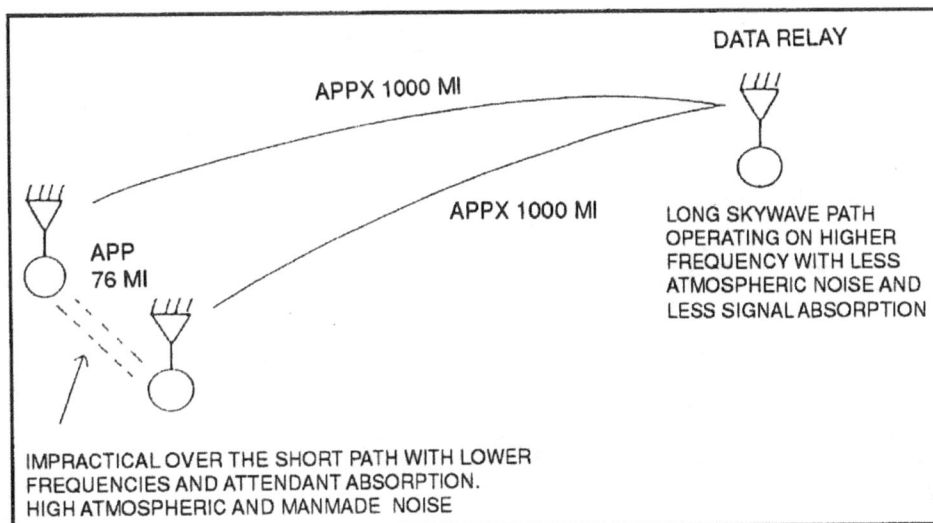

Figure 12. A means of improving path reliability by employing a long path relay as an alternative to a short path (primarily for data communications)

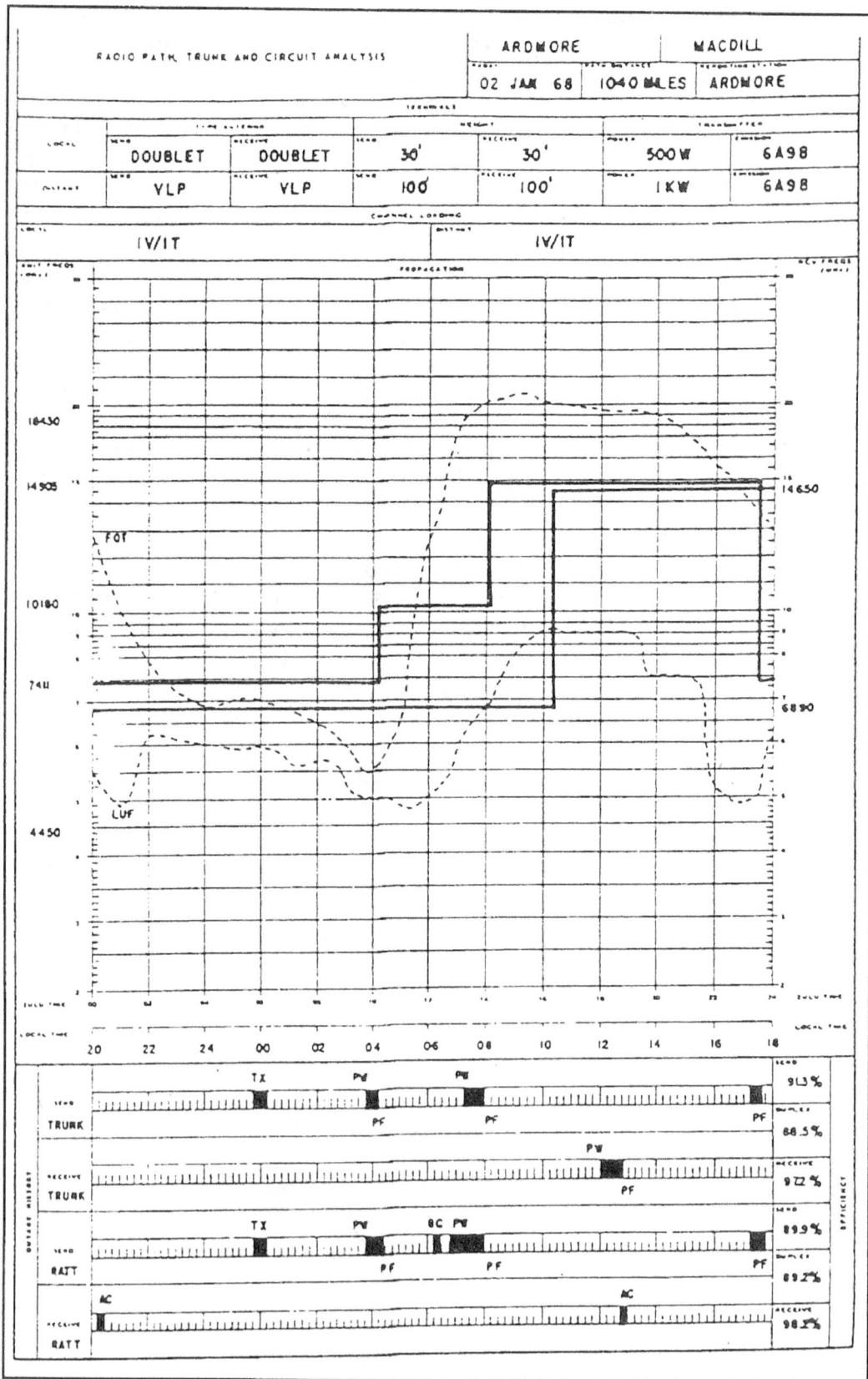

Figure 13. Analysis Worksheet for a RADAY, showing outage and propagation relation-ships.

Refer to Figure 13.

The short HF skywave path/near vertical incidence (NVIS) path (same thing — different names) is one of the more difficult tactical HF operations of all. Tactical HF radio operation is usually low powered, antennas are usually in efficient, noise levels are high, interference can be brutal, and assignment of an adequate frequency family is often a problem. There are techniques to help maximize the performance of the short haul HF skywave system, primarily by using the best available antennas, ensuring that operators are trained, using propagation predictions, employing proper equipment, and obtaining and effectively employing an effective radio frequency family. While it is sometimes impossible to provide effective communications over an HF NVIS path for each entire RADAY, it is possible to minimize outage and prepare users for the potential outage periods. The proper engineering and effective operation of the radio system are keys to success.

Near-vertical-incidence skywave (NVIS) propagation: the Soviet approach

by Lt. Col. David M. Fiedler

The poor performance of our HF-RATT and HF-SSB voice equipment in supporting fast moving, widely dispersed operations is not only the result of inadequate training and doctrine but of inadequate equipment.

In two previous articles in ARMY COMMUNICATOR, I have advocated the use of the near-vertical-incidence skywave (NVIS) for communicating beyond groundwave range—up to a distance of 400 km. With the NVIS technique, energy is radiated at a low enough frequency so that it is reflected back to earth at all angles by the ionosphere. This results in the energy striking the earth in an omnidirectional pattern without dead spots (without a skip zone) if an efficient short-path antenna such as a doublet is used. I had hoped that the Signal Corps would use this technique to solve some serious operational problems. I am still hoping. As a further argument, I would like to point out in this article that the Soviet Union has already seen fit to incorporate NVIS technology into their communications doctrine. Can we afford not to?

Historical background

Due to the huge size of the USSR and the problems they encountered in establishing land line systems over vast distances with sparse population, long range radio circuits early-on became an attractive alternative to the Soviet military. This, coupled with the military situation during and after the Russian Revolution in which large land areas were constantly being contested, gave Soviet communicators a great impetus into HF radio systems just at the time when early HF technology was providing equipment and techniques that would do the job. Soviet reliance on HF systems has continued to this day, and their capabilities include very good mobile tactical communications and over the horizon HF RADAR applications.

On the other hand, with the advent of satellite communications in the 1950s and 1960s, we in the west believed that HF radio systems had lost their military potential. Studies were conducted that concluded that satellite systems with their high reliability, huge bandwidths, and tremendous channel capacity would eventually replace all HF systems, even at the lowest tactical levels. This led to a virtual halt in HF equipment development in the 1960s and 1970s and is the reason why our forces still use equipment such as the AN/GRC-46, AN/GRC-26D, AN/GRC-122/142, and AN/GRC-106. It is also the reason why until recently most of our technical and doctrinal literature bore dates from the early 50s, and why the training of operators deteriorated to a very marginal level. The Soviets, who had our resources in neither space nor electronics at this time, and who did not share our belief in the invulnerability of satellites (with good reason since they were developing a satellite destroying weapons system), continued to develop HF radio technology.

The Soviets and NVIS today

Soviet Lt. Col. V. Natetov, writing in the military journal, *Tekhnika I Vooruzheniye (Technology and Armament)* No. 11 - 1985, outlined the Soviet view of NVIS training, operation, and doctrine. He also discussed a mobile NVIS capability not presently available to U.S. forces. According to Natetov, "Radio network operations are usually set up for communication over short distances (up to 300 km). Non-directional antennas or those with poor directivity and Zenith Radiation [the Soviet name for NVIS] should be used in this case. The most widely used antennas for this kind of communications are the horizontal or slanted symmetrical dipoles. For symmetrical dipoles there are also frequency limitations of use depending on operating conditions.

Therefore, this shortwave radio antenna set usually includes no fewer than two dipoles to establish communications in the different sectors of the set's band of frequencies. Generally speaking, in communicating at ranges up to 300 km, horizontal dipoles can be oriented arbitrarily in the area; however, it is best to set up the dipole perpendicular to the direction of the most remote correspondent."

What he says up to here tracks item for item with the data presented in my previous articles on HF communications. The more significant and disturbing part of Natetov's paper came later when he went on to say, "Ionospheric communication in motion and during brief stops at distances up to 200-300 km is conducted using Zenith radiating and receiving antennas arranged on top of the operating vehicle." Though this mobile NVIS antenna is a capability unheard of in the U.S. Army,* it is not surprising that the Soviets, who are the masters of mobile/mechanized war and HF radio communication, would combine the two if possible.

Natetov's article raises serious questions about the ability of the U.S. Army to provide its mobile forces with communications support comparable to that of the Soviet Army. (Interestingly, on page 24 of the Fall 1983 issue of AC, in an article by Maj. Charles H. Hill III, there is a photo of a Soviet BTR-60 displaying a unique "railing-like antenna array." I believe that this array is in fact an NVIS mobile antenna, exactly like the one Natetov describes, further evidence that the Soviets do indeed have the

*No unit that I know of except the New Jersey Army National Guard even attempts to use a crude form of mobile NVIS. The NJARNG tries to use whip antennas bent at a 45 degree angle in order to get some useable vertical skywave radiation and is presently experimenting with another NVIS mobile antenna.

hardware necessary to communicate with NVIS while on the move.)

This glimpse of the Soviet view of NVIS has led me to conclude the following:

• We are on the right track when advocating the use of NVIS using dipoles located close (.1-.25 wavelengths) to the earth for communication beyond groundwave range. Our communications doctrine and training must incorporate NVIS, and the reluctance of certain portions of the U.S. Signal community to change their thinking must be overcome.

• We have fallen short in supporting the combat/mechanized mobile force. The poor performance of our HF-RATT and HF-SSB voice equipment in supporting fast moving, widely dispersed operations is not only the result of inadequate training and

doctrine but of inadequate equipment. We need, and we must develop with all deliberate speed, a mobile NVIS capability along with the training and doctrine to support it. If we don't, tactical commanders will be tied to line-of-sight communications (HF and VHF) and area systems, which will not respond adequately to high-mobility battle situations.

Since the technology is known, a mobile NVIS capability can be developed and deployed quickly. I call upon the Signal Corps to recognize the military potential of mobile NVIS techniques—as the Soviets have—and make the effort to develop the necessary equipment and training to use it.

A summary of Col. Fiedler's background appears on p. 20.

Mobile NVIS: the New Jersey Army National Guard approach

Because of the size, shape, and vertical direction of radiation, communications equipment can be hidden in depressions and under cover, thus making it harder to find . . . Mobile NVIS will make possible the selection of much more survivable sites than those used today.

by Lt. Col. David M. Fiedler

In previous articles in the ARMY COMMUNICATOR on the subject of short path (0-400km), high frequency (HF), skip zone free, radio communications, I challenged Army training and doctrine while attempting to inform tactical communicators of a more efficient way to use their HF radio equipment.

In the Winter/Spring 1987 issue, I argued that the Soviet Union has a mobile near-vertical-incidence skywave (NVIS) capability that we lack, but one which we need if we are to support fast moving, deep penetration operations over wide areas.

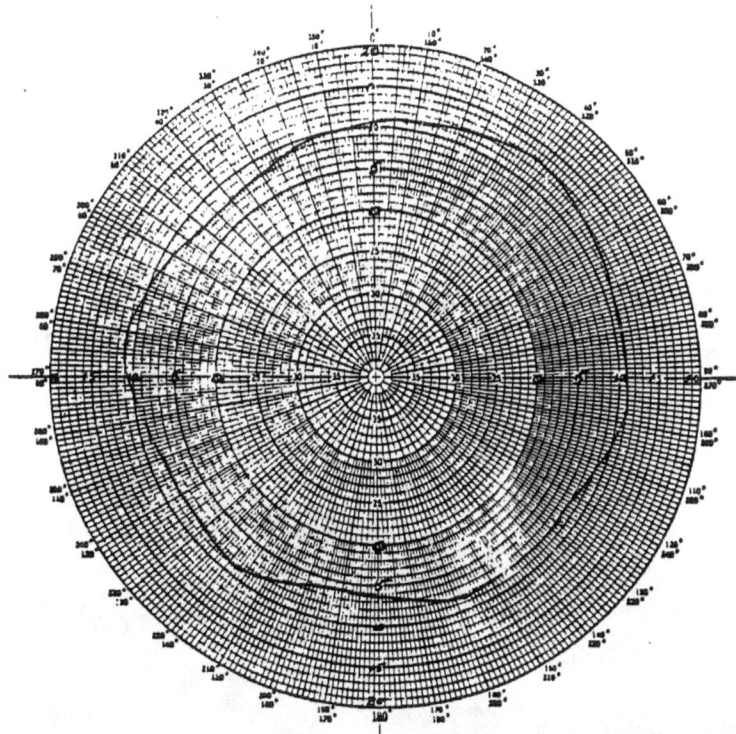

Figure 1. This NVIS antenna pattern is omni-directional, which means that the energy radiated from the antenna is of equal strength in all directions from the antenna. Such a pattern is achieved by radiating the signal in a near vertical direction at a frequency low enough for it to be reflected by the ionosphere. The effect is similar to that achieved by directing a water hose straight up at a flat surface: the water, hitting the surface above, rains down in a circular pattern without dry spots, covering the area below. All radio stations within the pattern will receive the same signal at approximately the same strength without gaps in coverage or "skip zones."

Since then, fixed operations using wire dipoles have been incorporated into FM 24-18, Appendix N, though the reader should still be careful when using this FM, since some incorrect data about short path propagation and "skip zones" still remain in other sections of the manual. Appendix N is the correct picture for short skywave paths.

In the area of mobile operations, the situation becomes more complicated. While it was easy to adopt standard issue wire dipoles (like the AN/GRA-50) for fixed-station NVIS operation, the Signal Corps' lack of interest in short skywave path communications produced no comparable mobile zenith radiating antenna or even the

requirement for one. This being the case, my search for a starting point to match the Soviet capability had to begin outside the Signal community.

Fortunately, the U.S. Army Avionics Research and Development Activity (AVRADA) was faced with a similar wide-area continuous-coverage communications problem when attempting to communicate with aircraft engaged in "nap of the earth" (NOE) flight. After consulting with Mr. John Brune and Mr. Frank Cansellor of AVRADA, I obtained Tech Report ECOM 4366 (Reference 1), which showed excellent omni-directional antenna patterns with aircraft-mounted transline antennas.

I felt that similar results could be obtained when the antenna was mounted atop tracked or wheeled vehicles as the Soviets were doing.

There is currently no TRADOC (Signal Corps) generated requirement for this capability. However, the New Jersey State Area Command (NJSTARC) does have a requirement for continuous communication over an operational area very similar to that occupied by a typical U.S. corps. With the STARC mission in mind, I was able to convince the commander and staff of the New Jersey National Guard (NJNG) of the value of this capability and receive their support. Unfortunately, the adjutant general of New Jersey has no funds or

Figure 2. Series AV-600 NVIS antenna installed on Army UH-1

Figure 3. Mobile NVIS antenna installed on NJNG communications facility

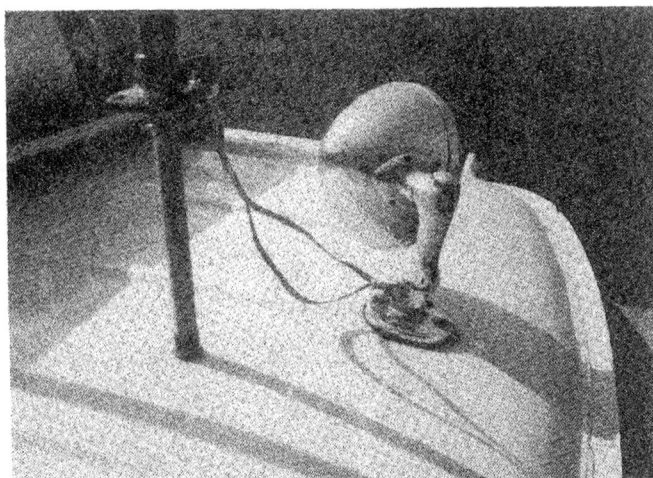

Figure 4. End grounded to searchlight mount

manpower for R&D projects, so the work had to be accomplished by NJNG personnel on their own time and without funds.

This being the situation, I contacted Mr. Seymour Greenspan, former chief of the Airborne Systems Technical Area of AVRADA, who was working with the company suppling AVRADA with their aircraft antennas. Mr. Greenspan, Mr. Florenio Regala, and Mr. Frank Hoar of TRIVEC-AVANT then arranged for the NJNG to get a series AV-600 shorted loop NVIS antenna in exchange for a copy of any operational test results obtained.

The antenna was mounted on the NJSTARC mobile communications facility shown in Figure 3. It was fastened to the cargo rails on the van roof for support and grounded on one end to the searchlight mount as shown in Figure 4. The antenna was fed from the other end by a Kenwood Model 4305 transceiver modified for military operations and matched by a Kenwood Model AT-230 antenna matcher using a single-wire feed line as illustrated in Figures 5 and 6.

Matching the antenna was the most difficult technical problem, since this type of arrangement causes the antenna impedance characteristics to be highly inductive. In order to make the antenna match and thus radiate the maximum amount of signal toward the zenith in a vertical pattern similar to Figure 7, the matching unit

had to be sufficiently capacative to cancel this inductance. In the beginning of the effort, we thought that the AT-230 matching unit did not have sufficient capacitance and that more would have to be added either at the feed end or the ground end of the antenna. Fortunately, we found that when tuning the AT-230 with the band switch in the 10MHz band, the unit had sufficient capacity to match the antenna at the operational frequencies of NVIS (2-8 MHz). This method is probably not as efficient as using a matcher designed specifically for the TRIVEC antenna, but it was simple and sufficient to do the job with the equipment on hand.

After completing the installation as shown, we conducted operational tests. Since we lacked the equipment necessary to measure either the actual antenna pattern or the amount of gain toward the zenith, much of our testing had to be of a less scientific, but more practical nature.

First, we established contact with fixed stations from the mobile facility at various ranges and azimuths. Each fixed station was equipped with the radio equipment described above, as well as with a horizontal dipole antenna .1-.25 wavelengths high, using an operating frequency of 4520 kHz. The locations of the stations and their range from the mobile

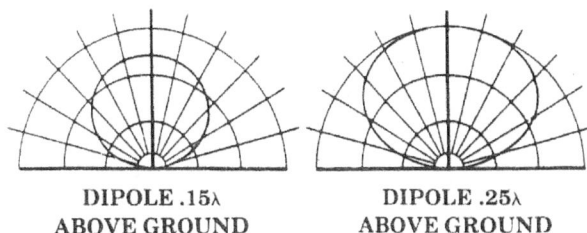

DIPOLE .15λ
ABOVE GROUND

DIPOLE .25λ
ABOVE GROUND

Figure 7. In order to achieve the omni-directional antenna pattern shown in Figure 1, the transmitter must radiate in a vertical direction as shown. This is achieved by locating a horizontal antenna .1-.25 wavelength above the ground so energy will be reflected at a high angle. This energy is in turn reflected by the ionosphere toward the earth, creating the circular area of coverage shown in Figure 1.

Figure 5. HF transceiver and antenna matching unit

Figure 6. Antenna single wire feed point

facility are shown below and in Figure 8 (net station map). The performance was compared to a standard vertical whip antenna used on the same mobile facility, whose groundwave range, when communicating to the same fixed stations, never exceeded 25-30 miles. Results are shown below.

Station	Range	Conditions
Cape May NJ	79mi	beyond groundwave range (flat)
Dover NJ	48mi	beyond groundwave in (mountains)
East Orange NJ	47mi	beyond groundwave (urban rolling)
Jersey City NJ	51mi	beyond groundwave (urban rolling)
Lawrenceville NJ	.2mi	groundwave (flat)
Long Branch NJ	41mi	beyond groundwave (flat)
Morristown NJ	43mi	beyond groundwave (hills)
Newark NJ	47mi	beyond groundwave (hills)
Phillipsburg NJ	40mi	beyond groundwave (hills)
Plainfield NJ	32mi	beyond groundwave (hills)
Red Bank NJ	36mi	beyond groundwave (flat)
Riverdale NJ	58mi	beyond groundwave (hills)
Sea Girt NJ	38mi	beyond groundwave (flat)
Somerset NJ	34mi	beyond groundwave (rolling)
Teaneck NJ	60mi	beyond groundwave (hills)
Vineland NJ	53mi	beyond groundwave (flat)
Westfield NJ	33mi	beyond groundwave (hills)
West Orange NJ	49mi	beyond groundwave (hills)
Woodbridge NJ	36mi	beyond groundwave (rolling)
Woodbury NJ	.34mi	beyond groundwave (flat)
Bordentown NJ	10mi	groundwave (flat)
Fort Dix NJ	15mi	groundwave (flat)
Fort Drum NY	270mi	beyond groundwave
West Orange NJ	44mi	beyond groundwave (hills)

Communications were established with all of the above stations on 4520 kHz during the days of the tests. All

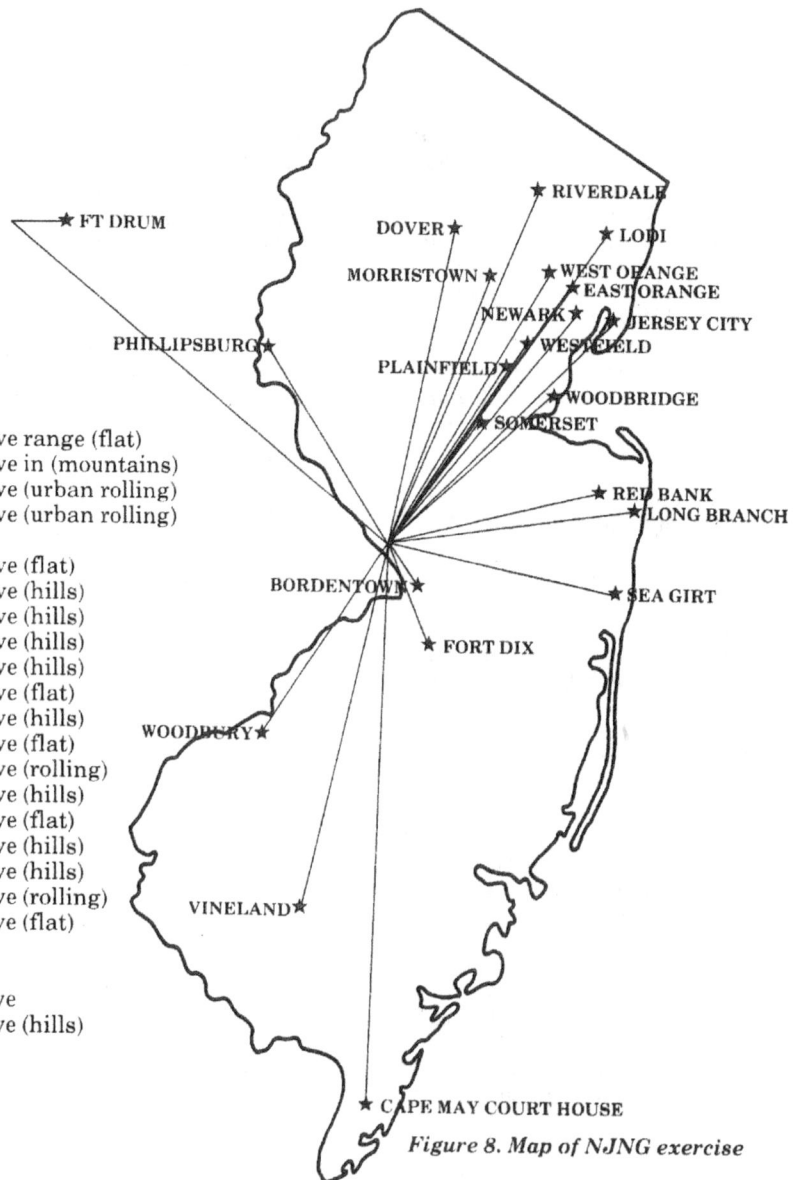

Figure 8. Map of NJNG exercise

stations could communicate with the mobile facility, and the mobile facility could communicate with all stations.

Second, we drove the mobile facility and communicated with fixed stations (in this case NJSTARC at Trenton and the NJNG base at Bordentown). We did this in the hill-covered areas of Mercer County, N.J., at ranges between 0 and 20 miles, purposely picking locations that were hidden behind terrain features so that groundwave communications were not possible. These tests were conducted both standing and on the move. Military considerations, such as cover and concealment, were also considered when picking locations. Good command post, but normally poor communications locations (when using current non-NVIS techniques) were purposely selected. Periodically during the tests, communications checks were also conducted with Fort Drum, N.Y., (range 260 mi) and West Orange (range 42 mi). At no time was the mobile facility, whether fixed or on the move, unable to contact either Fort Drum or West Orange.

Our third test was conducted to determine if, in fact, the antenna was directing the bulk of its radiated power in a vertical direction as shown in Figure 7. This was done to confirm that the mobile facility and its excellent communications results were indeed a product of NVIS design. We used an ME-31 field strength meter to detect the radiated field strength of the TRIVEC antenna from the ground level to a height of 30 feet above and next to the van. The meter showed no deflection at ground level but deflected steadily upward as it was elevated to approximately 2/3 scale at 30 feet elevation. While this is not an instrumented antenna range type of test, it does confirm that NVIS as a ground mobile concept is valid.

When constructing this system, we observed all the rules of frequency selection for NVIS systems explained in my previous papers and confirmed by the Soviet journals. Daytime operational frequency was below 8 MHz, and antenna height was as close to .1 wavelength as the height of the van would permit. In addition, antenna feed lines were kept as short as possible, and particular care was taken when grounding the end of the antenna to the van to assure a good ground connection.

The results of this work are very clear. We can, if we use NVIS techniques, communicate with all units in an area larger than that typically occupied by a U.S. corps. We can do this for both voice and data modes of communications, without gaps in coverage and while engaged in high mobility, Deep Battle/Deep Attack operations. In addition to improving communications, using mobile NVIS also results in the following significant military advantages:

• Electronic warfare: Since all radiated energy returns to earth from above at approximately the same signal strength, direction finding on the signal becomes very difficult, and the probability of intercept and detection is greatly reduced.

• OPSEC: Because of the size, shape, and vertical direction of radiation, communications equipment can be hidden in depressions and under cover, thus making it harder to find. In fact, the criteria for selection of HF radio communications sites will have to be revised, because mobile NVIS will make possible the selection of much more survivable sites than those used today.

• Physical detection: Studies by the Armor Center have shown that often the first item detected on a vehicle with the engine off is the vertical radio antenna. An NVIS antenna is flat and much harder to detect.

• Safety: Horizontal mobile antennas do not have a spear-like construction, which in the past has caused injury and even death to U.S. troops.

If there is any doubt in anyone's mind that mobile NVIS works, let me take this opportunity to invite them to New Jersey to see for themselves. Let me also take this opportunity to again urge the Signal Corps not only to teach this technique properly as part of a carefully designed program of instruction, but also to create the necessary requirements so that we can build on this preliminary work in the area of mobile operations and eventually catch up to the Soviet mobile capability. We know it works; we can see it in operation. Why should the NJNG and the Soviet Army be the only ones who can make it pay off in combat?

I wish to publicly thank the following for their help in this effort: Maj.Gen. Francis R. Gerard, Brig.Gen. Kenneth L. Reith, and Col. William Singleton, the adjutant general, deputy adjutant general, and chief of staff, respectively, of the New Jersey National Guard, for their command support of this effort; Brig.Gen. William Harmon, program manager, Joint Tactical Fusion Program, who provided encouragement when no one else wanted to listen; Mr. Seymour Greenspan, Mr. Florenio Regala and Mr. Frank Hoar of the TRIVEC-AVANT Corp., who provided the antenna hardware for a good cause and at a good price (gratis); and CWO Robert Herka and CSM Thomas Hannon of the NJNG, who provided the labor to put it all together in their spare time and also put up with my many phone calls to ask them if they were done yet.

References

1. U.S. Army Electronics Command, Research and Development Technical Report ECOM-4366, Avionics Laboratory, November 1975, by John F. Brune and Joseph E. Reilly.
2. **The Rules of the Antenna Game: What Every Ham Must Know About Antennas,** W5WJR Antenna Products, P.O. Box 334, Melbourne, Fla. 32902-0334 by Ted Hart.
3. TRIVEC-AVANT Corp. drawing D 60188 (AV 605-()), rev A TRIVEC-AVANT, 17831 Jamestown Lane, Huntington Beach, Calif. 92647.
4. U.S Army Field Manual FM 24-18, **Field Radio Techniques,** December 1984, Appendix N.
5. "Beyond line-of-sight propagation modes and antennas," by David M. Fielder and George Hagn, ARMY COMMUNICATOR, Fort Gordon, Ga., Fall 1983.
6. "Skip the skip zone: we created it we can eliminate it," by David M. Fiedler, ARMY COMMUNICATOR, Fort Gordon, Ga., Spring 1986.

Lt. Col. Fiedler, a member of the National Guard, is the author of several articles on tactical communications and electronic warfare. He has served in Regular Army and National Guard Signal, infantry, and armor units in both CONUS and Vietnam. He holds degrees in physics and engineering and an advanced degree in industrial management.

Lt. Col. Fiedler is presently employed as the chief of the Fort Monmouth Field Office of the Joint Tactical Fusion Program, and as the assistant project manager for Intelligence Digital Message Terminals. He is also the director of systems integration for the JTFP. Concurrently, he is the chief of the C-E Division of the New Jersey State Area Command, NJARNG. Prior to coming up to the JTFP, Lt. Col. Fiedler served as an engineer with the Army Avionics, EW, and CSTA Laboratories, the Communications Systems Agency, the PM-MSE, and the PM-SINCGARS.

8: OPTIMIZING LOW POWER HIGH FREQUENCY RADIO PERFORMANCE FOR TACTICAL OPERATIONS

by Lt. Col. David M. Fiedler

These digital techniques combined with the antenna techniques are the optimum limits of our present knowledge. On the battlefield they will provide the critical difference IF WE IMPLEMENT THEM.

Ever since the Fall of 1983 when ARMY COMMUNICATOR (AC) published its first article on the subject of Near Vertical Incidence Skywave (NVIS) antennas and propagation (written by George Hagn and myself), I have received dozens of positive phone calls from Regular Army, Army National Guard, and Marine Corps communicators on the subject of how successful vertical HF radiation is for tactical communication over multi-Corps size areas at ranges of from 0 to 600 miles.

In each and every case, when the antennas and frequencies were properly selected in accordance with the techniques described in my basic (Fall 1983) and subsequent papers, the results were exactly as shown in the papers. The rate of reported success using properly matched (or cut), correctly erected antennas and standard 400 watt radios such as the AN/GRC-106 or the AN/GRC-193 has been virtually 100%. Use of lower power radios in the 75-150 watt range has also shown similar results. Unfortunately, users of very low power (20 watt) radios such as the AN/PRC-104 (manpack) or the AN/GRC-213 have not achieved this high a level of success, even when using good antenna and frequency selection techniques and assuring the maximum radiated power that the radio is capable of is being radiated. NVIS Communications under average conditions using this equipment can be achieved reliably; however, performance under degraded conditions is definitely below that of the higher power radios even when all components have been properly selected and installed.

The explanation for this is not difficult and is best shown by example: When using a modern radio receiver, we can expect to receive and process signals entering the radio at a level of approximately -110 decibels

(db) provided the level of the signal is at least 10 db above the ambient noise level (Noise level at the receiver is the sum of all noise: noise received with the signal, noise receiver generated noise, and so on). Therefore, if the noise level is -120 db and a radio signal is received at a level of -110 db, we can receive this signal (at the lowest receiver sensitivity level of -110 db) and communicate. The signal level at the receiver is calculated by adding the transmitter power, the antenna gain (or loss) at both the transmitter and receiver and the path (propagation) loss. If we have an ideal situation using a 20 watt radio (36 db) and an antenna with no loss (i.e., cut wire dipole) and a path loss of -110 db (typical of losses in free space), the received signal will be at -74 db which is 36 db above the minimum required signal level, and we can communicate very nicely assuming the average noise level stays at -120 db, or less.

Unfortunately, it is not unusual for noise levels to jump to -90 db or more due to atmospherics, locations with high levels of man-made noise such as urban areas. Under these conditions, the received signal level now required to communicate becomes -80 db (10 db above noise level), and what was a 36 db signal margin now becomes only a 6 db margin. At the same time, path losses can vary up to 10 db depending on path length and atmospherics. Additionally, most antennas used for tactical communications are not perfect and have some (if small) loss of gain associated with them (see Table 1). Under these conditions, probability of communications with a 20 watt radio becomes less due to the lower power level. The 6 db reserve margin may get used up resulting in loss of communications. The most obvious solution to this problem is, of course, to radiate more transmitter power—go for example from 20 watts (36 db) to 400 watts (47 db) and gain 11 db which will put us back into a good position for communication. Unfortunately, since we are concerned in this case with radios that have a fixed (and low) output power level, this is not an option.

Since the path losses and atmospherics are not under our control, the only area we can turn to in order to squeeze out every last db of gain and thus increase our probability of communication is the antenna. At this point, it is important to remember two things: first, every 3 db of gain obtained by improving the antenna is equivalent to doubling the transmitter power, and, second, we are assuming a correct frequency selection in the range of frequencies that will be reflected from the ionosphere (as described in the original paper) is selected. This frequency is usually in the 2-4 MHz range at night and the 4-8 MHz range during the day.

Table 1 shows the typical gain towards the Zenith for various common antennas. As you can see, depending on the antenna selected, gain can vary from 3 db (same effect as doubling the transmitter power) for the Shirley folded dipole (see Figure 1) to -41.5 db for a whip antenna with very little power radiated vertically.

The Shirley dipole (Figure 1), while it does have the most gain at the correct angles for NVIS communication, is more than a bit cumbersome to erect and is not easily transported. The next best antenna shown is the simple half wave unbalanced single wire dipole. This antenna is cheap (approximately 31 dollars and up) and only requires the erection of two masts whose optimum height is 1/4 wavelength above ground at the operating frequency. It gives acceptable performance at heights as low as 15 feet. Figure 2 is a plot of vertical power gain vs frequency for dipoles at various heights above the ground. If one follows the NVIS frequency selection rule of thumb shown in the 1983 paper (2-4 MHz daytime, 4-8 MHz nighttime), this figure shows that an antenna height of 30 feet (over good ground) is optimal for coverage of the NVIS band of frequencies, without having to adjust antenna heights.

In addition to height above the ground, the physical shape of the dipole also has an effect on vertical gain. Any wire dipole fed from the center and not supported will sag in the center due to the weight of the

Antennas	Clearing		75-ft forest clearing		50-ft forest clearing
h/2 unbalanced single-wire dipole	+1.0	-2.8	0.0	-1.2, -1.70.0	
h/2 balanced single-wire dipole	+0.5	-3.7	no data	no data	no data
h/2 folded dipole (30-0:5-0 U balun)	+0.2	-1.0	no data	no data	no data
h/4 short (loaded to h/2) dipole	-3.0	-5.2	no data	no data	no data
h/2 sleeve dipole (on ground)	-32.1	-28.3	no data	no data	no data
3-freq. fan dipole @ 15 ft	-0.4	-5.1	no data	no data	no data
3-freq. fan dipole @ 12	-2.4	-5.6	no data	no data	no data
3-freq. fan dipole @ 9	-4.0	-8.1	no data	no data	no data
shirley folded dipole	+3.0	-0/3	no data	no data	no data
3 h/4 inverted L (1:h = 2:1)	-0.0	-2.8	no data	no data	no data
3 h/4 inverted L (1:h = 3:1)	-0.8	-3.3	no data	no data	no data
3 h/4 inverted L (1:h = 4:1)	-1.0	-5.8	no data	no data	no data
3 h/4 inverted L (1:h = 5:1)	-2.0	-6.3	-10.2	-10.7, -12.5	-9.0
30° slant wire (h/r elevated)	-10.1	-14.8	-11.8	-13.5, -14.2	-14.0
60° slant wire (h/r elevated)	-11.8	-14.8	no data	no data	no data
10-ft square (vertical plane) loop @ 6 ft	-24.1	-25.3	no data	no data	no data
16.5-ft whip	-41.5	-44.0	-31.7	-25.0, -25.2	no data

Summary of relative gain toward the Zenith for Field-expedient HF antennas

Table 1.

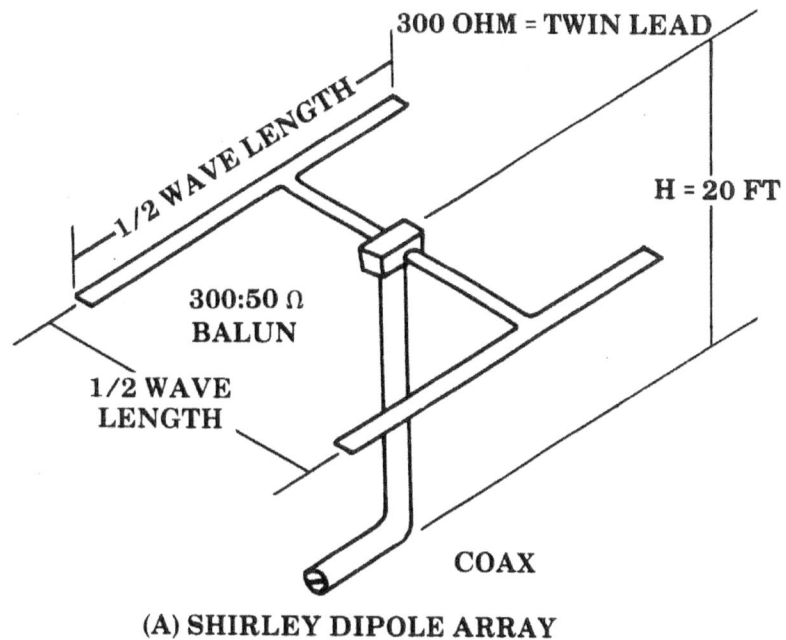

Figure 1. The Shirley Dipole

transmission line and feed connector. Figure 3 shows the antenna pattern plot of a dipole that sags 5 feet is horizontal and is supported 5 feet above the horizontal. It shows that the sagging dipole has up to 3/4 db more of vertical gain over a supported dipole (center 5 feet above ends) and

even more over a horizontal dipole. The extra power toward the Zenith was not created by any magic. The total amount of transmitter power is constant. What the change in dipole geometry has done is shift power from low angle radiation (not useful for short paths) to the higher angles useful for the NVIS mode of transmission. By allowing a sag in the dipole of 5-10 feet we can gain up to 1.6 db over a horizontal antenna. This isn't all that much, but it's free, and in marginal, low power situations when every db counts, it's certainly better than nothing. In terms of cost effectiveness, letting the dipole sag can increase effective radiated power by 50% without having to buy a power amplifier, and that's not bad.

Additional gain can be obtained by use of the antenna shown in Figure 4. This dipole adds an extra wire that acts as a reflector approximately .15 wave lengths below and 5% longer than the active (radiating) dipole. In theory, this arrangement can add 5-6 db of vertical gain to a dipole effectively providing 4 times the effective radiated power without having to buy anything more than a few dollars worth of wire and insulation. The height above ground of the reflector (lower) wire is not critical; however, separation of the two antenna wires should be as close to .15 wave length as possible. For example, if the antenna is 30 feet high as selected from Figure 2, the reflector should be 16-18 feet below it for best results.

If we raise the center of a dipole while keeping the ends at constant height, we form an inverted V. This antenna also radiates well in the vertical direction (see Figure 5) and has the advantage that only one mast needs to be erected. The critical thing about this antenna is that as the center is raised to form the inverted V, the apex angle decreases. If the antenna is analyzed and the currents are broken into vertical and horizontal components, it can be seen that the current in one leg is in the opposite direction from the current in the other leg. With this geometry, the radiation associated with these currents tends to cancel. An inverted V often has an apex angle of 90 degrees which does not give as good a vertical gain as V's with apex angles of 120 degrees or more. As can be seen in Figure 5, to get the most vertical gain from an inverted V antenna apex angles of 120 to 140 degrees must be maintained. This is done by raising the ends of the antenna while keeping the center constant. When using the inverted V, the reflector techniques shown in Figure 4 for the dipole is also valid. The combination of proper apex angle (120-140 degrees)

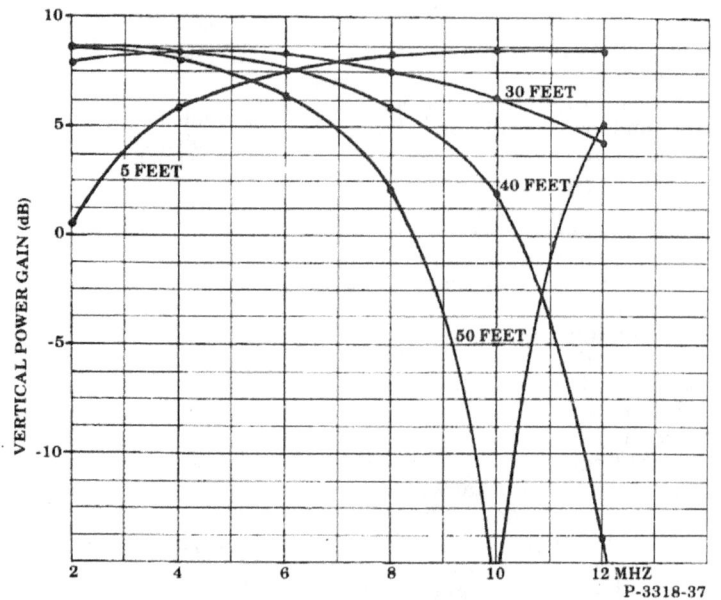

CUT 1/2 λ DIPOLE AT VARIOUS HEIGHTS OVER PERFECT GROUND

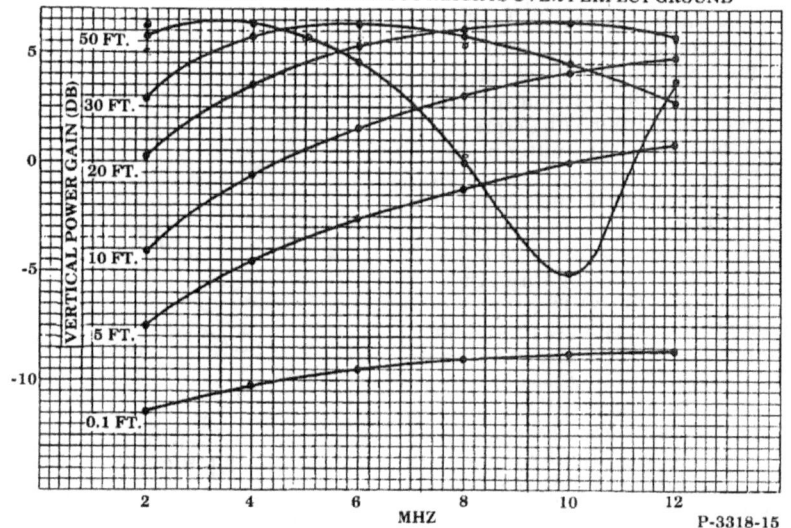

Figure 2. CUT 1/2 λ DIPOLE AT VARIOUS HEIGHTS OVER AVERAGE GROUND

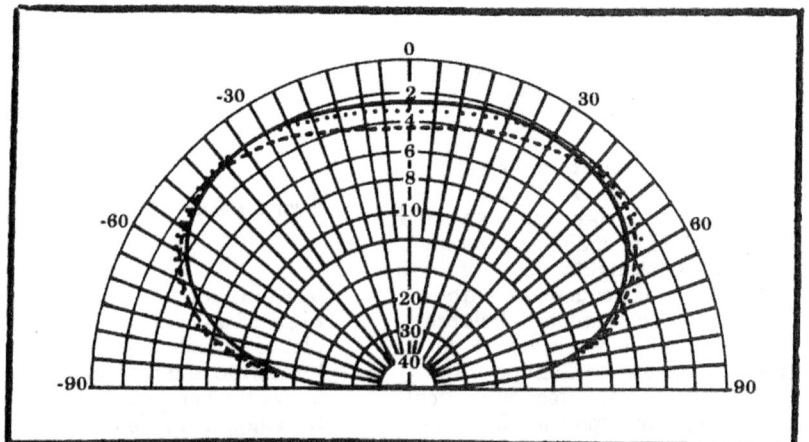

Figure 3. Typical dipole performance from ref 1 top line 5 foot sag, third line horizontal dipole, second line dipole center supported 5 feet above ends. Increment lines represent variations of 2 db.

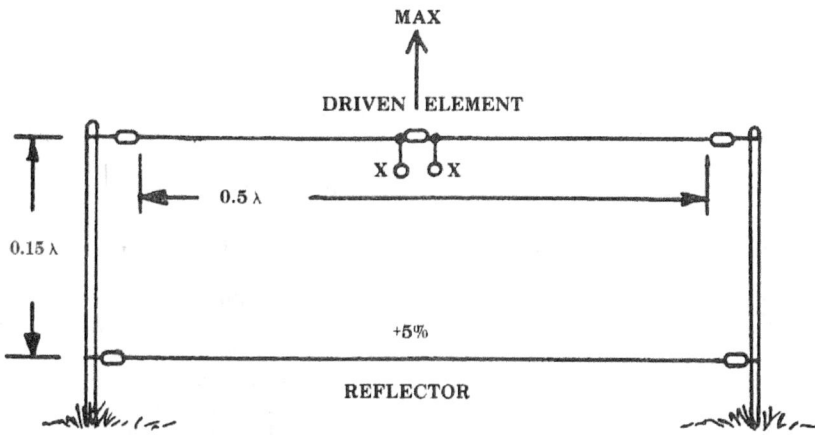

Figure 4. Half wavelength dipole with reflector

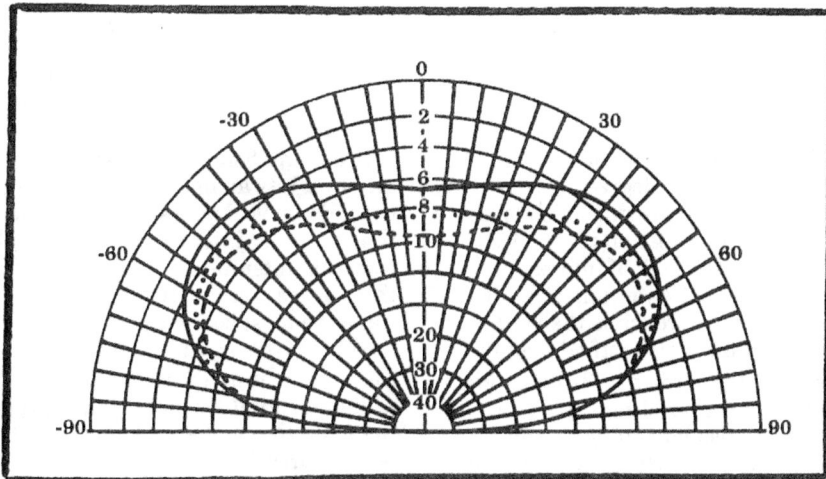

Figure 5. Typical inverted V performance from ref 1 top line apex angle 120 degrees next line apex angle 100 degrees next line 90 degrees (ends of V are elevated). Increment lines represent variations of 2 db.

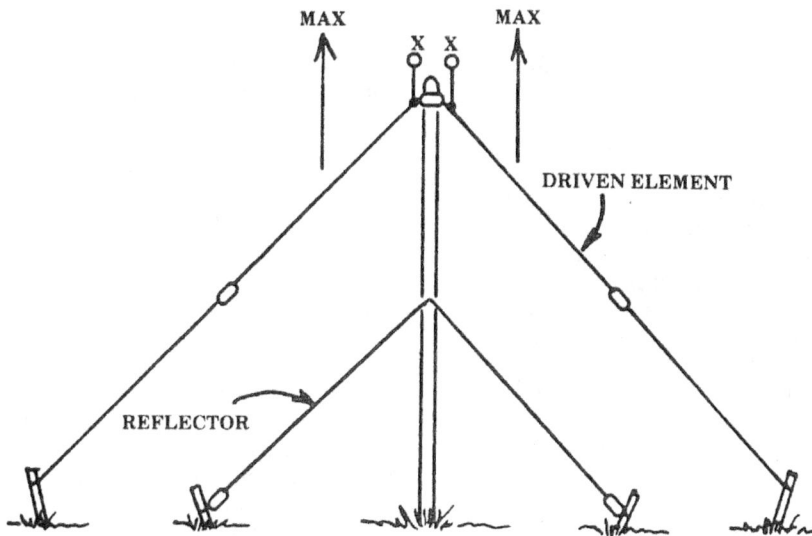

Figure 6. Inverted V antenna with reflector

and wire reflector .15 wave length below the active radiator will give good vertical gain characteristics and the convenience of having to erect only 1 mast as shown in Figure 6. Another good vertical radiating antenna is shown in Figure 7. This antenna is called a full wave loop since each side is 1/4 wave length and it is constructed parallel to the earth. At the design frequency, the radiation is straight up and reliable communication at ranges out to 600 miles is possible. The antenna is also very broad band and requires very little matching. The optimum height for this antenna is the same as a dipole (.25 wave length), but there are two significant problems that must be dealt with: in order to properly match this antenna and feed it from a standard military radio using a co-axial transmission line, a 4:1 balun transformer is required to provide impedance matching and to isolate the antenna from the transmission line and prevent pattern distortion due to interaction between antenna and transmission line, and the antenna itself requires a full wave length of wire. At the lower frequencies this can amount to several hundred feet of wire.

In summary, there are several very good and very cheap ground mounted antennas and installation techniques that when used in conjunction with low power radio sets will provide better performance under poor conditions over the 0-600 mile range. By simply letting an antenna sag or by adding an additional wire, we can effectively increase radiated power by a significant amount. In some cases, the amount is equivalent to gains produced by increasing transmitter power 2 to 4 times. Under marginal condition, where path losses, jamming, natural or man-made noise, weak transmitter batteries, and so on are effecting communications and every db of gain counts, these methods can make the difference between accomplishing our mission or not. From a cost effective point of view, these methods will reduce the need for power amplifiers while giving equal performance at almost no additional cost. From the tactical point of view, by proper selection of antennas and frequencies forces using HF radios, we can carry smaller radios that require less support items (power supplies, batteries, etc.) and still get the job done.

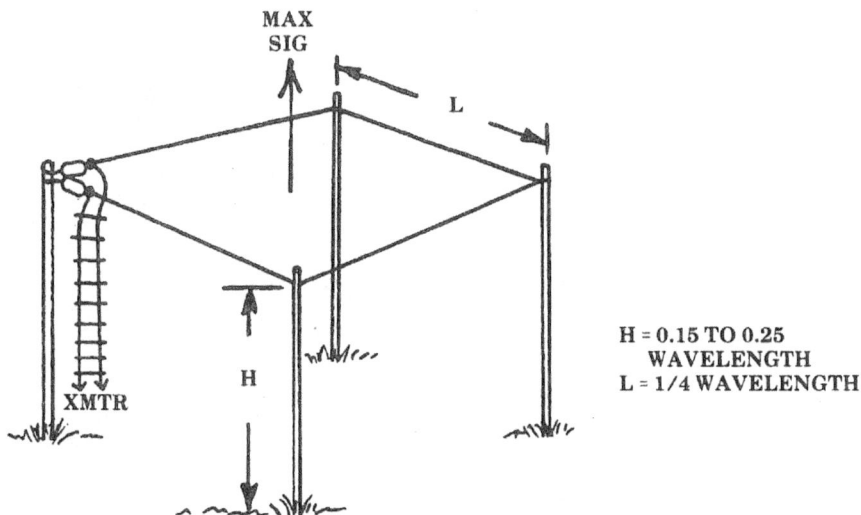

MAX
SIG

L

H = 0.15 TO 0.25
WAVELENGTH
L = 1/4 WAVELENGTH

XMTR

H

Figure 7. Full wave loop antenna. Note if coaxial cable is used a 4:1 BALUN transformer is required.

This discussion so far has primarily addressed techniques to improve communication using ground mounted low powered HF manpack radios such as the AN/PRC-104. Let us now turn to the vehicle mounted situation and the same type low power radio, i.e., the AN/GRC-213.

In the Fall of 1987, AC published a paper of mine on mobile NVIS antenna. In that paper, I showed the results of a test using a fore-aft mounted vehicle loop antenna constructed by the New Jersey National Guard and built from an antenna currently being used on Army helicopters. This test proved that omni-directional skip zone free coverage out to several hundred miles could be easily and effectively achieved by this type loop antenna— in this case the vehicle itself formed part of the loop—if the loop could be resonated (matched). Further research has shown that if the antenna (loop) circumference length is between 1/8 and 1/3 wave length at the operating frequency, radiation efficiencies approaching that of a dipole can be achieved. The Soviets have fielded several of these type antennas which are detailed in AC Winter/Spring 1987 issue, page 16. If this type antenna is adopted—and right now indications are that the Army is NOT moving in that direction (even though we should)—the loop will provide the best all around antenna for low power radios in tactical mobile applications. Elevation plane patterns for the fore-aft loop are shown in Figure 8. Note

the excellent gain towards the Zenith between 50 and 90 degrees.

Since loops for the time being seem to be out of the question for the US Army due to the lack of interest at both the Signal Center and CECOM, the next best approach is the bent whip antenna. This method simply takes the standard 15 or 32 foot whip (made longer by adding extra whip sections if possible) antenna and bends it in the horizontal direction as flat as possible either away from or over the vehicle as shown in Figures 9 and 10. Bending the whip over the vehicle forms the equivalent of an asymmetrical open-wire line which does have radiation in the vertical direction but significantly less than the same whip bent back as flat as it can be bent, or a fore-aft loop. Whip antennas bent backward (away from the vehicle) as flat as possible form the equivalent of an asymmetrical dipole. This configuration approaches the efficiency of a loop with a significant amount of gain in the 60-90 degree elevation range.

There are problems with both bent whips and loops on vehicles that must be addressed in order to get every db of NVIS gain possible for mobile operation. They are:

• In order to get the maximum radiated power, the loop (or any antenna) must have the impedance of the antenna matched to the transmission line and the transmitter. Since the Army was ignorant to the advantages of NVIS propagation for many years, some standard antenna

matchers cannot match a loop antenna below about 4 MHz since they do not have sufficient capacitive reactance. In those sets of equipment that have this problem, additional capacitance must be added in order to properly match the antenna. This is not a big problem and can be easily accomplished at the lowest (operator) level by a small applique box once the proper values of the capacitive component are determined for the equipment in use. Fortunately, the CU-2064 antenna coupler used in the AN/GRC-193 (IHFR) will match both loop and whip antennas without modification.

• In order to get vertical radiation from the whip antenna, it must be pulled flat either forward or backward. This operation causes two problems: First, the antenna mounts are made to keep whip antennas vertical and bend only when the antenna hits an obstruction. Pulling them down 90 degrees is an attitude they were not intended for. This tends to break antenna mounts, quickly at their bases. Second, whip antennas have the most current density at the feed point (base). This is the part of the whip that produces the most radiated signal and is also the part of the whip that is the least horizontal (see Figures 9 and 10) when the whip is bent. This attitude produces a pattern that is off center and reduces gain at the desired angles. This problem also has a solution. A whip tilt adaptor (see Figure 11) developed by Allan Christinsin of the Air Force solves the problem by allowing the entire whip including the base to lay flat. This gives maximum gain towards the Zenith and the best whip operation.

In summary, vehicular, low power NVIS antennas give varying amounts of gain in the following order (best to worst) providing proper antenna matching is done:

• Fore-aft loop.
• Horizontal whip using whip tilt adaptor so whip base is flat (directed either fore or aft).
• Whip bent back as flat as possible without breaking but at least 45 degrees (asymmetrical dipole).
• Whip bent forward as flat as possible without breaking (open wire line).

In addition to these techniques, a whip can be made into a loop for better performance if it is long enough

3 MHz

7 MHz

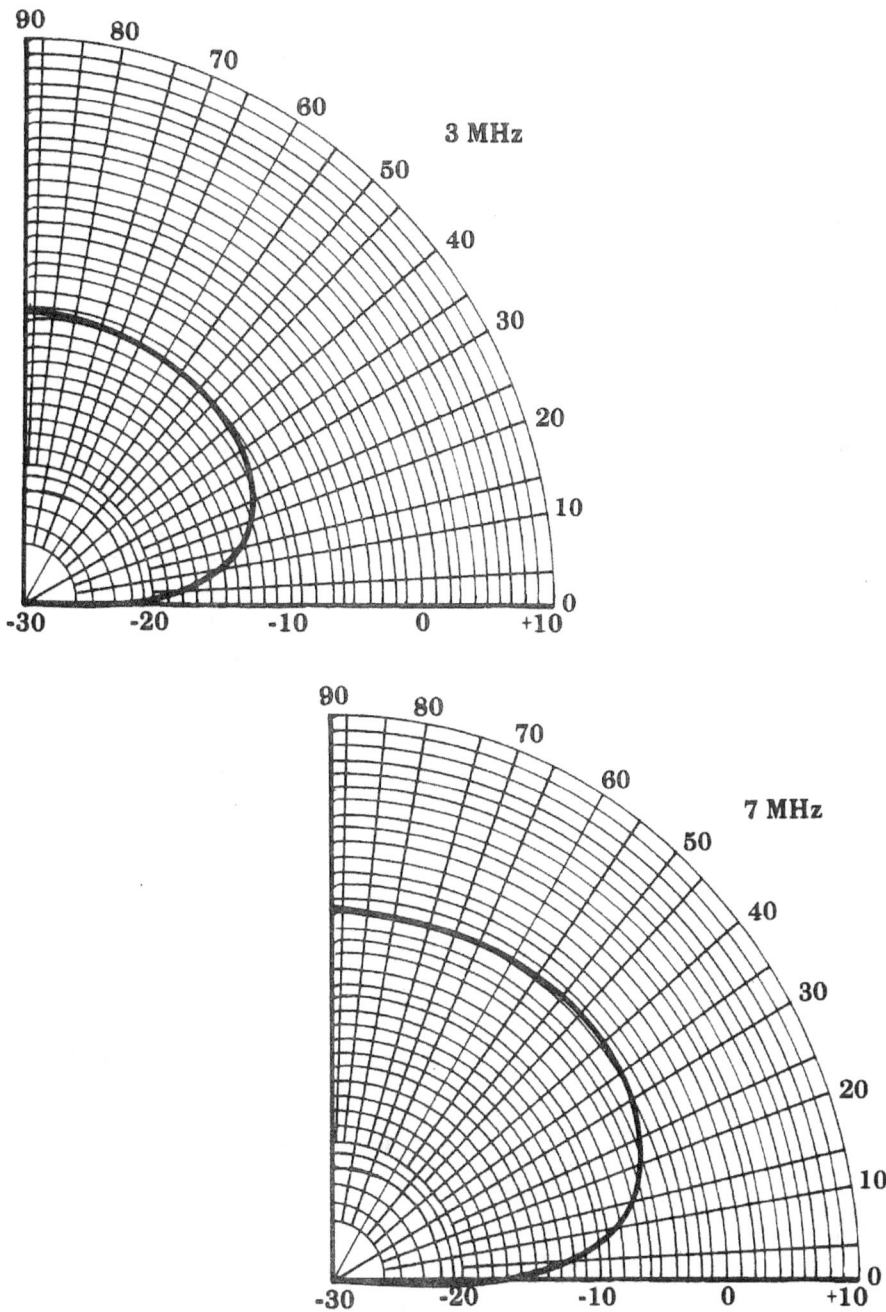

Figure 8. Elevation plane pattern for the fore-aft mounted loop

to be grounded to the front of the vehicle. This technique has been tried by US forces in Korea by attaching a side mount battery connector to the whip tip and screwing the terminal lug into the vehicle front bumper to ground the whip and make a loop using the vehicle as part of the antenna as shown in the New Jersey National Guard paper. It should also be noted that in most cases for grounded whip antennas, longer is better, with the total circumference of

the loop (including the vehicle) being between 1/8 and 1/3 wavelengths for best performance.

My purpose in writing this paper is twofold. The first being to impart to my fellow tactical communicators some more and improved techniques that can be used to attain better gap free NVIS HF communications over multi-Corps areas with equipment now in the field. My second is to present yet another plea to both the Signal Center and CECOM to provide

the user with all the equipment and information needed to do the HF communication job right. The updating of FM 24-18 with the information in my first paper certainly helped improve the HF communication picture in the field (judging from my many phone calls). What is needed now is some specific simple to produce hardware and supporting information in order to get more performance out of our equipment, such as:

• Additional wire and insulator in the standard AN/GRA-50 antenna kit in order to make a higher vertical gain antenna as shown in Figure 4.

• Loop antennas (fore and aft) for wheeled and tracked vehicles to match the capability already on Army Aircraft, and deployed by Soviet forces.

• Whip tilt adaptors so that the bent whip technique can be used effectively, where loops are not provided, or cannot be matched.

These techniques have all been proven by actual operations by the New Jersey National Guard, US Forces Korea, and several Marine Corps units. The costs to implement them through the force are minimal (in some cases the cost of 30 yards of wire). Not to make these improvements when we can is certainly a disservice to our troops and the commanders we support. The time to act has long since passed, so we must act NOW to correct our deficiencies before combat operations show us the error of our ways.

The antenna techniques described above are the best possible combination of technical and tactical factors available to squeeze all the gain possible out of HF radios and antennas. They will all work well (and exactly the same way) with either low power or high power applied, but for low power applications the extra gain margins attained can prove to be the critical difference between communicating and not.

Once these techniques are applied, we will have optimized the voice (analog) radio portion of a tactical HF system. There is, however, one more source of gain available. This is the so called "processing gain" achieved by going from voice communications to digital data communications.

When using data terminal devices such as the standard AN/PSC-2 Digital Communication Terminal (see

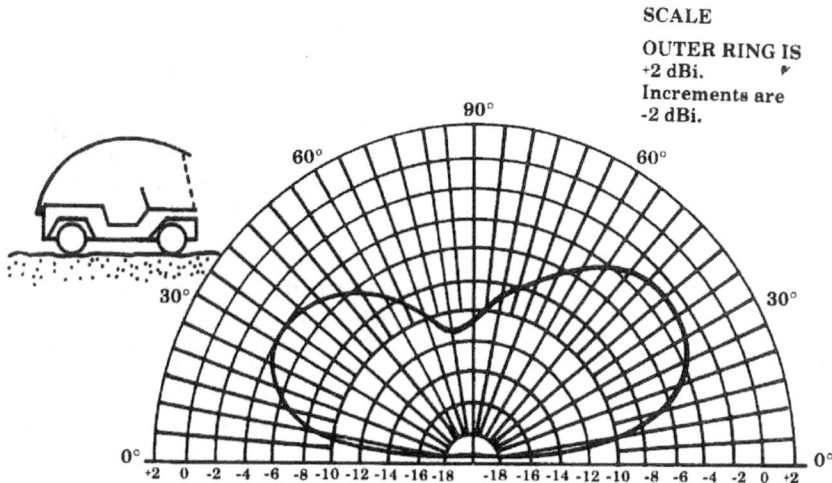

Figure 9. Far-field elevation pattern of vehicular 15 ft. whip tied to front of vehicle at 10 MHz.

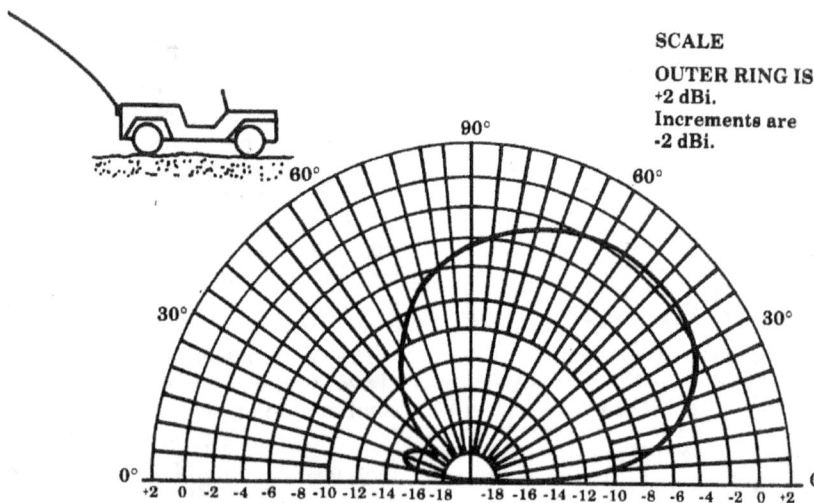

SCALE

OUTER RING IS
+2 dBi.
Increments are
-2 dBi.

Figure 10. Far-field elevation pattern of vehicular 15 ft. whip bent backwards at 45° at 10 MHz.

of the receiving terminal several times. Correct messages can be created from several incorrect transmissions by automatically merging the good data packets from several incorrect transmissions to form a good message. Due to the short transmission time required to the ability to merge data packets, and to powerful error detection and correction coding technique data, signals that are very close to the noise level and near the limits of receiver sensitivity level can be recovered. In radio terms, where analog voice signals need to be typically 10 db above the noise level for communications to occur on HF radio, data signals can be much lower (sometimes as much as 6 db lower) and data communication will still work. This difference called "processing gain" is the equivalent of transmitting at a higher power level. These digital techniques combined with the antenna techniques are the optimum limits of our present knowledge. On the battlefield they will provide the critical difference IF WE IMPLEMENT THEM.

I would like to thank Maj. Gen. Francis R. Gerard, the Adjutant General of New Jersey and Brig. Gen. Kenneth Reith, the Deputy Adjutant General along with Brig. Gen.(P) William E. Harmon, Program Manager of the Joint Tactical Fusion Program for their encouragement to continue my efforts in improving tactical HF communication for the total force. Hopefully, these gentlemen will be paid off in the form of better tactical communications for both the NJNG and the JTFP. I would also like to thank Allan

AC Winter/Spring 1987), packets of digital information are transmitted, received, checked, error corrected and displayed in a very small fraction of the time required to send a voice message over the same media (in this case low power HF radio).

When data is sent from the terminal via a self contained Modulator/Demodulator (MODEM), the MODEM output level can be adjusted to assure the proper degree of modulation, and the maximum radiated signal is present at all times.

In addition, the rate of the data transmitted can be adjusted to compensate for the effects of such common HF radio problems as high noise levels, bursts of noise, selective fading, multi-path, etc. At all speeds of transmission, the data is separated into packets and coded for error detection and correction. Data bits are then spread in time for immunity to net noise. In addition to this, data packets received with errors that cannot be corrected by the coding are automatically repeated at the request

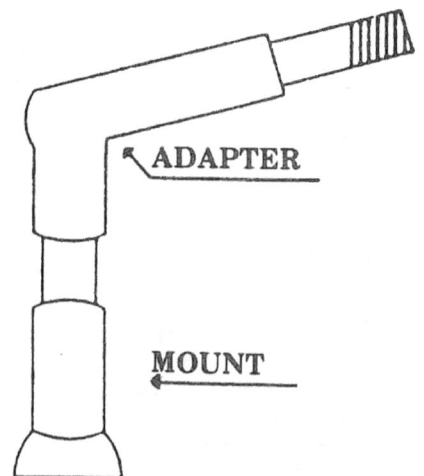

Figure 11. Whip tilt adapter

Christinsin for the details of his whip tilt adaptor. I hope it soon becomes standard equipment for all services. I would also like to thank Bob Jacobson, Jerry Neal, and Gabe Luhowy who provided invaluable technical information. Most importantly, I would like to thank the many Army and Marine Corps field users who have called me to report their successes with NVIS techniques during actual operations, their encouragement kept me going and their desire to accomplish their missions has motivated me greatly.

References

"American Radio Relay League," **QST Magazine,** April 1984, "The Effect of Real Ground on Antennas," p. 34, James Ravtio.

"American Radio Relay League," **W1FB's Antenna Notebook,** chapter 5, Doug DeMaw.

US Air Force, **The Tactical Communicator,** HQ TCD, Langley AFB, Va., p. 15, "Air Support Operations Center HF Communications, A. S. Christinsin.

US Army CECOM Technical Report 80-C-0580, "Tactical Antenna System Technology Assessment," Collins Communications Systems Division, Rockwell Corporation.

ARMY COMMUNICATOR, Fall 1983, "Beyond line-of-sight propagation modes and antennas," by David M. Fiedler and George Hagn.

ARMY COMMUNICATOR, SPRING 1986, "Skip the 'skip zone'; we created it and we can eliminate it," by David M. Fielder.

ARMY COMMUNICATOR, Winter/Spring 1987, NVIS, "The Soviet approach," by David M. Fiedler.

ARMY COMMUNICATOR, Fall 1987, "Mobile NVIS: the New Jersey National Guard approach," by David M. Fiedler.

Lt. Col. Fiedler has served in Regular Army and National Guard Signal , Infantry, and Armor units in both CONUS and Vietnam. He holds degrees in physics and engineering and an advanced degree in industrial management.

Fiedler is presently employed as the chief of the Fort Monmouth Field Office of the Joint Tactical Fusion Program, and as the assistant project manager for Intelligence Digital Message Terminals. He is also the director of systems integration for the JTFP. Concurrently, he is the chief of the C-E Division of the New Jersey State Area Command, NJARNG. Prior to coming up to the JTFP, Lt. Col. Fiedler served as an engineer with the Army Avionics, EW, and CSTA Laboratories, the Communications Systems Agency, the PM-MSE, and the PM-SINCGARS.

Command and control:
HF radio communications and high angle antenna techniques

by LTC David M. Fiedler (NJARNG)

Over the past several years, I have tried my best to impress upon AC readers the military utility of tactical HF radio operating over extended ranges using high angle skywave propagation techniques (Near Vertical Incidence Skywave—NVIS).

Based upon the most recent information coming from SWA on the subject, it has been proven again in actual combat operations over corps and division size areas that the horizontal antennas and frequency selection techniques I recommended in my previous AC papers work well when using medium power radios (100-400 watts) and fixed station or vehicular antennas.

Also highly successful, if somewhat less effective, tactical HF radio communications over these areas was achieved when the radio equipment used consisted of a medium powered base station and a small low powered (20-50W) out station, such as the AN/PRC-104 or AN/GRC-213. These configurations are commonly found in SOF, LRSU, and calvary scout applications. The reduction in

Techniques outlined here will help greatly in overcoming the adverse effects of terrain, weather and enemy jamming.

effectiveness was, of course, due to several factors such as:

• Low power radio sets transmit less powerful signals making them harder to receive than higher power stations.

• Manpack sets use battery power sources which produce weaker signals as the battery is used up.

• Low power manpack and vehicular radio sets are equipped mainly with 10-15 foot vertical whip antennas, which are inefficient energy radiators due to their electrically short length. Whips also produce energy at low radiation angles unsuitable for the ranges desired (see previous AC papers). The vertical whip, though it is issued with all tactical HF radio sets, is in fact the

poorest antenna for short range beyond-line-of-sight (BLOS) HF communications (30-400) miles.

Knowing these factors, we can easily improve system performance using low power manpack sets. Some techniques are:

• Operators must be trained to make sure a fully charged battery is used whenever possible in order to ensure full performance from the radio, particularly the transmitter.

• Whip antennas supplied with all HF radio sets must be replaced in BLOS applications by horizontal wire antennas, such as simple long wires, dipoles, inverted "Vs," and so on.

These antennas all produce the high angle radiation needed to communicate over longer distances. Due to their physical length, these antennas are also more efficient radiators of electrical energy.

As shown in Figure 1, when a wire antenna, such as a dipole of proper length, is elevated, gain at high angles increases. One should remember that every 3db of antenna gain improvement is equivalent to doubling the transmitters power.

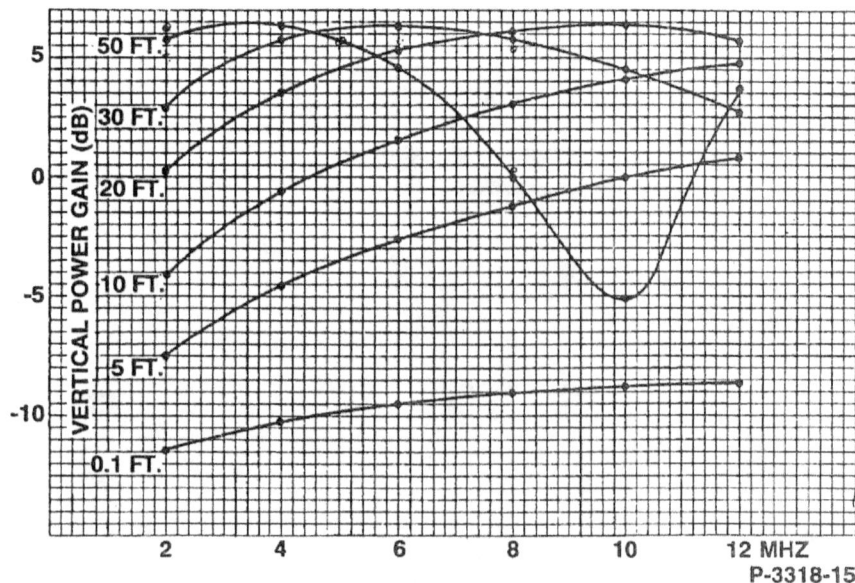

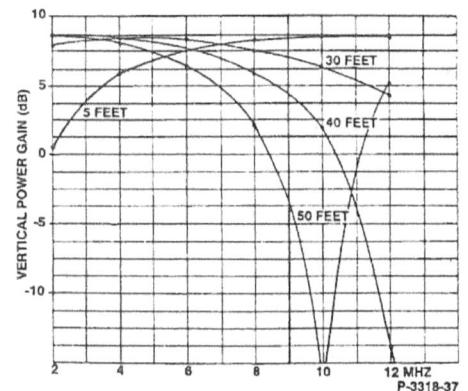

Figure 1. Cut 1/2 λ dipole at various heights over perfect ground.

Even a simple 30-50 foot length of wire placed horizontally (see figure 2) along or on the ground will produce a high angle pattern with much improved gain (when compared to the whip) at the near vertical angles required to cover corps and division size areas. Care must be taken to assure that the length of the wire is within the range of the radios antenna coupler so that the maximum amount of radio energy possible is transferred to the antenna for transmission. The radio TMs should be consulted to determine the optimum wire length for a particular radio and frequency.

In the event that this information is not available, a field strength meter (ME-61) or a VSWR meter (AN/URM-182) can be used to determine the best match of radio to antenna length.

In any event, as long as the wire does not get higher than .25 wavelength or longer than .5 wavelength, the basic high take off angle omnidirectional pattern shown in figure 3 will be created. Gain will vary with antenna geometry; however, this pattern will allow good communications out to 400 miles using a wide range of wire lengths and heights provided that an operating frequency that does not exceed the maximum useable frequency (MUF) for the situation is selected.

Wire antennas can be erected part vertical and part horizontal or in a sloping configuration. Almost any wire support will work. Even installations where the antenna wire is draped over vegetation or hung out

of building windows will work and will always out perform whip antennas. The wire antenna does not need to be in a straight line; the most important factor is to try to keep the radiating wire as high and as clear as possible.

• Use a counterpoise. A counterpoise is a grouping of radial wires located on the ground under the radio and its antenna. The counterpoise is connected to the radio ground terminal or to chasis ground from a central point in the counterpoise. The counterpoise forms a high capacity path to ground from the antenna. This reduces the negative influence of the earth on antenna efficiency by reducing ground currents and making available more energy in the antenna for signal radiation.

The reduction in ground currents lowers system losses and increases effective radiated power so signal strength improves.

Practical counterpoises for manpack applications can consist of an "X" made of two eight-foot lengths of stranded wire (test lead wire works well) joined at the center with solder or a hose clamp, or an electrician's "bug nut" or "U" bolt and a small wire running from the center to the radio ground terminal. The radio just sits atop the center point of the "X" (see figure 2B).

This simple modification has been shown in some cases to improve signal power by 3-5db for a radio sitting on the ground. A ground stake should

also be inserted if possible, but this is not a must. More and longer wire radials will also yield improvement, but the effect is smaller each time a radial is added. The counterpoise technique can also be used while manpacking by taking about four feet of computer ribbon cable and shorting both ends. One end is then connected to the radio ground terminal, and the free end is allowed to trail down along the chasis. This ribbon trail will reduce ground currents and produce more effective radiated power in the same way as a full counterpoise. The manpack configuration is the worst possible configuration for good communications.

If the radio is allowed to remain on the operator's body, the ground is capacitively coupled through the body. In this situation, almost all of the radio's output power is absorbed by the high loss ground system. This is the major reason for the poor performance of manpack radios. Simply placing the radio directly on the ground will give a measureable improvement in efficiency and effective radiated power.

It has to be recognized that cover and concealment considerations may not allow a small unit such as LRSU or SOF detachment to erect a very high or very long wire antenna or counterpoise; however, as figure 1 shows, a difference of just a few feet in elevation can make a good deal of difference in effective radiated power.

The best compromise between OPSEC and communications at the small detachment (manpack) end of a

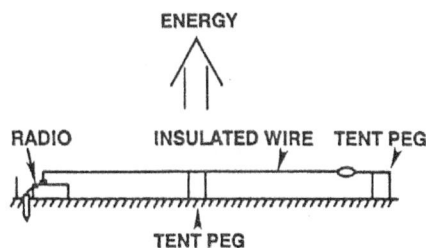

Figure 2. Simple wire antenna for low power radio. (Height can be from a few inches to 30 feet. Do not let wire contact ground. Use more pegs to keep wire off ground if needed.)

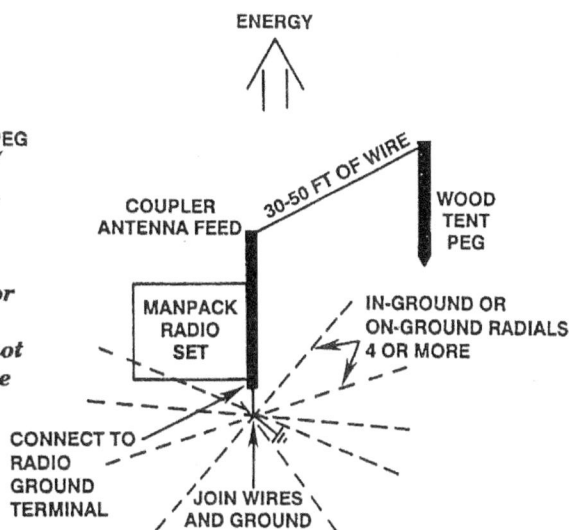

Figure 2A. Manpack radio with counterpoise to increase effective radiated power.

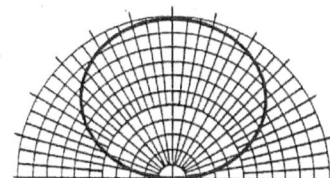

Figure 3. Typical vertical radiation pattern for horizontal antenna. Note: Energy at the high angles (35-90 degrees) suitable for communications over 0-400 mile paths.

radio circuit is usually a simple 30-40 feet of straight wire antenna placed 1-5 feet above ground using tent pegs, wood poles or the like, and a counterpoise under the radio. This combination gives the high take off angle needed with a reasonable amount of energy being radiated from the low power radio set.

In recent years, several antenna manufacturers have attempted to sell the Army "throw on the ground" multi-wire antenna configurations that claim to solve both the OPSEC problem and improve communications. Despite these exaggerated claims, none of these complicated "spider webs" that take up areas of up to 20 x 150 feet have produced significantly better performance than a simple straight wire tent-peg high or so off the ground. These type antennas and their claims should be discarded since they are tactically not very practical.

Once we have taken these measures to make the manpack station the best that we can make it under the operational situation, we must then turn to the base station end of the circuit in order to significantly improve the overall system's performance.

To begin with, the base station will usually have a higher power transmitter (100-400 watts) which will help improve the system in itself because higher power transmitted signals will be easier to detect at the out station receiver.

Additionally, because it is a base station, there is usually more leeway with antenna size and less movement involved. This will allow us to construct the most efficient antenna possible at the base station and, therefore, improve the overall system

reliability—while at the same time achieving a reasonable trade-off with tactical considerations at the out station.

Of course, if the situation permits and if both ends of the systme can have similar high efficiency antennas, we will be at the optimum level for communications. This holds true even if both ends of the system are using low power radios because of a significant amount of system improvement is due to the use of the more efficient antenna, not a more powerful transmitter.

As can be seen from figure 1, the half wave horizontal dipole gives the best gain at the right takeoff angles (35-90 degrees) for NVIS communication when kept at about a constant 30 feet off the ground. Height adjustments are really not required over the 2-10 MHz frequency band that will produce NVIS modes of propagation desired for 0-400 mile circuits.

There are, however, several antenna configurations that if installed properly can give better performance at the needed takeoff angles. Again, bear in mind that each 3db of antenna gain is equivalent to doubling the radiated transmitted power, and under poor propagation conditions or when jamming is present, this can often make the difference between success and failure. The best choices for the base station antenna follow.

• Horizontal dipole with reflector .1 to .25 wavelength below the radiating element (see figure 4). This antenna has a theoretical gain of 5.4db (the actual gain is probably somewhat less) above a horizontal dipole without reflector, and if configured as

shown, it provides both the gain and takeoff angles necessary for good communications.

This antenna can be constructed with one standard AN/GRA-50 antenna kit plus two extra insulators and an additional length of heavy wire to form the reflector. The antenna can also be configured as an inverted "vee," using one mast in order to save installation time; however, gain may be reduced slightly.

When using the "V" antenna configuration, remember to keep the apex angle between 120-140 degrees. Reflector wires are very useful when operating over poor ground since they greatly lower system losses; however, the distance between the radiating wire and the reflector does become the effective height of the antenna, so be careful with the wire spacing. This technique over good ground may not yield a dramatic gain at all since good ground has low system loss to begin with, but in areas like the Middle East or the NtC, the effect is worth the effort.

• The Shirley folded dipole. This antenna consists of two halfwave folded dipoles .5 to .65 wavelengths apart and 1/10th to 1/4 wavelength off the ground (see figure 5). This antenna is actually an array of 3 horizontal half wave dipoles whose energy is phased together (added) to form a more efficient high radiation angle antenna.

The Shirley does not pick up noise and interference from low angle sources, which helps system performance.

The antenna can be constructed out of TV twin lead but needs four masts

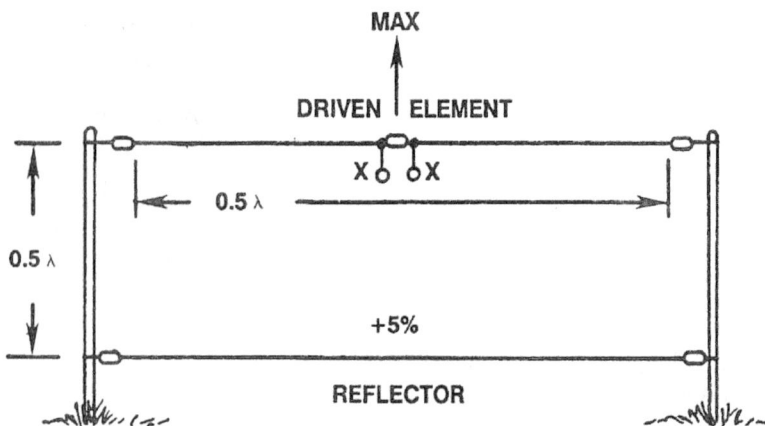

Figure 4. Half wavelength dipole with reflector

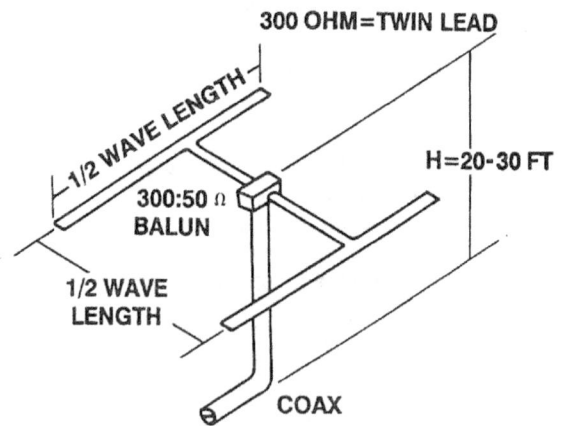

Figure 5. Shirley dipole array

and more space than the single wire dipole with reflector. It has a gain of approximately 3db over a single wire dipole since the electric fields from all three dipoles combine (add) to result in more gain at the high angles needed for NVIS type communications.

• A third very good base station type antenna very popular with the British Commonwealth Forces is the Jamaica antenna shown in figure 6. This antenna is a high efficiency, specialized antenna for high angle skywave (NVIS) use. Two full wave single wire dipoles parallel and 1/2 wavelength apart are needed. The distance between full wave dipoles is critical so that the signals in both dipoles will be properly phased together to form a composite (higher gain) signal.

The dipoles should be fed with two-inch open wire line (ladder line) connected to the radio with 75 ohm coax cable. Impedence matching is provided by the radio set antenna coupler.

The antenna is very efficient with gain towards the zenith of almost twice that of the Shirley antenna and four times that of the single wire dipole (without reflector). Because it is so directional toward the high angles, it does not pick up noise or interference from low angle sources including ground wave jammers.

The Jamaica can be constructed from two AN/GRA-50 kits, but it does need six masts and takes a while to construct. When trying to work with low power manpack stations, the combination of gain and low angle noise reduction at the base station provided by this type antenna will often make the difference between communications and no communications to the out stations.

The Jamaica is probably the most complicated tactical wire antenna that most organizations should try to construct out of common materials due to its large size and complexity.

It is important to remember that while these antennas can be used to replace expensive power amplifiers or give any station (including low power manpacks) more effective radiated power, they are all essentially single frequency resonant antennas—that is, they should all be constructed to the proper length for the operating frequency plus or minus a few percent in order to operate at peak efficiency.

Antenna couplers will help broaden the operating frequency range of the antenna, but separate day (high operating frequency) and a night (low operating frequency) antennas may be required depending upon the situation, frequency assignments, and the equipment used.

These type of higher gain antennas have proved themselves in the past for use in special operations and low intensity conflicts (LIC) when friendly forces were able to operate from secure base camp locations.

In the jungles and mountains of Malaya and Vietnam, base camp communications to widely separated and constantly moving patrols equipped with low power radio sets and horizontal antennas was maintained successfully by using this type antenna equipment at base locations and proper frequency selection techniques. In more modern applications, they can certainly be used in Central and South America for counter narcotic teams and in AirLand Operations applications for communicating from the support areas to MI, Scout, LRSU and SOF units deployed in the battle zone and/or the detection zone.

I urge commanders of these type units and all HF radio users in general to construct these type antennas and experiment with them. They will find that they will be able to communicate with high grades of service over wide areas and at ranges currently not thought possible by some units.

They will also (in some cases) eliminate the need for high powered radio sets and power amplifiers, thus saving equipment for other uses and cutting procurement costs.

These techniques will help greatly to overcome the adverse effect of terrain, weather and enemy jamming and should be incorporated into unit SOPs in order to assure the best possible HF radio circuits for command and control of our forces.

As we have seen in Southwest Asia, the fast moving pace of modern armored warfare will cause us to outrun all other means of terra based communications except properly engineered HF radio.

Similarly, in slow moving, low intensity conflicts (LIC) in mountain and jungle terrain, vegetation and weather will also render line-of-sight tactical radio systems including Mobile Subscriber Radio Terminals (MSRT) and SINCGARS ineffective except for relatively short range circuits.

This means that we must stress more than ever before, the proper use of HF radio and high angle antenna techniques on the battlefields of the future.

LTC Fiedler is a graduate of SOBC, SOAC, the Radio and Microwave Systems Engineering Course, and the Command and General Staff College. He has served in Regular Army and National Guard Signal, Infantry, and Armor units in both CONUS and Vietnam. He holds degrees in both physics and engineering, and an advanced degree in industrial management.

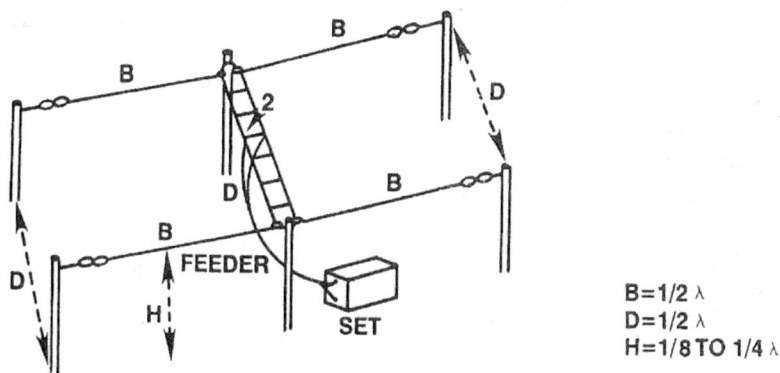

B=1/2 λ
D=1/2 λ
H=1/8 TO 1/4 λ

Figure 6. Jamaica antenna (Can be built from standard antenna kits AN/GRA-50; has four times the gain of the dipole antenna.)

How to survive long range and special operations

by LTC David M. Fiedler, NJARNG

In order to understand how a buried dipole works, we must first review what happens when radio frequency energy traveling in free space enters another medium--in this case, the earth.

One of the most dramatic lessons learned from the recent Gulf War was the great value of properly used Long Range Surveillance Units (LRSU) and Special Operation Forces (SOF).

In order to be effective, of course, these units must have good radio communications between the forward units and their operating bases. This is normally accomplished by means of manpack (low power) High Frequency (HF) radio or single channel satellite (SATCOM) radio. Unfortunately, other lessons learned in the Gulf were:

* There are usually not enough satellite channels and equipment available at the right time and place to provide SATCOM service to all the units we would like to employ, so HF radio must be used extensively.

* It is very difficult to conceal both HF and SATCOM antennas (both types are discovered with equal frequency).

* If good communications are not achieved, it is not worth sending a unit into danger.

* Poor antenna concealment leads to unit discovery and elimination.

At present, viable methods of concealing highly directional and sensitive SATCOM antennas are still being developed; there are, however, some existing proven and simple methods already available for concealing HF radio antennas and making them work. Of these, the construction of a subsurface (buried) antenna is the simplest and most effective.

Many types of subsurface antennas have been tried by US Forces in the past, but, as it turns out, a simple half-wave horizontal dipole or dipole array buried just deep enough for concealment is the best for LRSU/SOF hidden applications. There are no standard subsurface antennas in the Army inventory, so using units must construct them locally in the field, using available materials and standard components.

In order to understand how a buried dipole works, we must first review what happens when radio frequency energy traveling in free space enters another medium--in this case, the earth.

While subsurface dipoles produce some groundwave energy, high angle Nearly Vertical Incident Skywave (NVIS) is the dominant mode of HF radio communications for this application. For more detailed information on NVIS theory and techniques, see AC, Fall 1983 or Spring 1986 and FM 24-18.

If we look at Figure 1 which is a classic diagram found in many antenna text books, we can see that skywave energy arriving at an above ground antenna X from a distant transmitter is the sum of the skywave energy from the direct wave A and the reflected wave b. The energy from the reflected wave (b) arriving at antenna X is the energy of wave B reduced by some amount (N) because some of the energy is not reflected, but continues to travel on a bent path (C) into the earth.

The energy arriving at subsurface antenna S is what's left after wave B gives up reflected energy (B-N) and also loses energy to ground absorption (g), which is proportional to the depth that S is in the ground (d).

Received energy at the subsurface antenna S is, therefore, the skywave energy from B less the energy reflected b and the ground absorption loss g.

Remember, also, that the skywave energy has already been reduced at the point where it enters the ground by the path losses in free space and the reflection loss from the ionosphere (approximately 110db) total.

The bottom line is that for reception, the energy received at a subsurface dipole is considerably reduced by ground and free space path losses. Therefore, transmitter power should be as high as possible at the sending station in order to communicate reliably.

Also, receivers need to be as sensitive as possible, and correct selection of antenna(s) frequency is critical. The depth of the buried antenna should be as shallow as the tactical situation will allow for concealment of the antennas to reduce ground losses and emplacement time. This is particularly

important when the ground is wet or snow covered.

To transmit from a buried antenna, start at S, the energy becomes skywave and travels along B so there is no reflection loss (b), but there is ground absorption--which again can be reduced by keeping the antenna as near to the surface as possible.

Another problem when transmitting from a subsurface horizontal dipole is the problem of efficiency versus height above ground. As the energy field of antenna comes in contact with the earth, power is dissipated that would normally end up as radiation energy if the antenna was above the ground (see Fig 2). Even though the classical vertical pattern of the dipoles located close to the earth remains the same, when the antenna is subsurface (see Fig 3), the efficiency (radiated energy) is much less compared to an elevated antenna. Therefore, at least a two dipole array should be used to increase radiated power. Practically, this means that the station using the subsurface antenna needs the highest power transmitter and the best

antenna array and frequency selection possible to be successful.

Here is where equipment selection--AN/PRC-104 (20 watts) vs AN/PRC-132 (50 watts), antenna, and frequency may be critical.

In order to best communicate using a buried antenna, many studies (see references) have shown, that a half-wave center fed dipole (doublet) (see Fig 4), well insulated and surrounded with a good dielectric (such as air or teflon) and free at the ends, is the best.

The dipole is usually configured as a two antenna array for better performance. It is also known that the standard rule of thumb for calculating wavelength = 300/ (frequency in MHz) is no longer true since the effect of the ground is to increase the electrical "length" (decrease physical length) of the antenna because the velocity of propagation is slowed due to the effects of the earth.

Since both the earth's characteristics and the depth of the antenna will vary, it is very hard to calculate the proper length of a resonant antenna. In some cases,

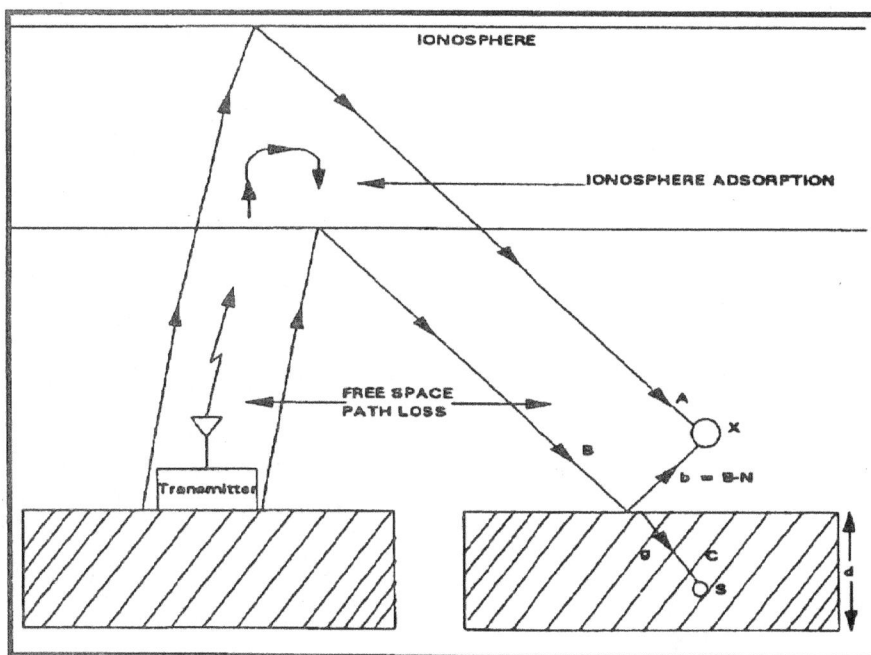

Figure 1 - Reception below ground via skywave

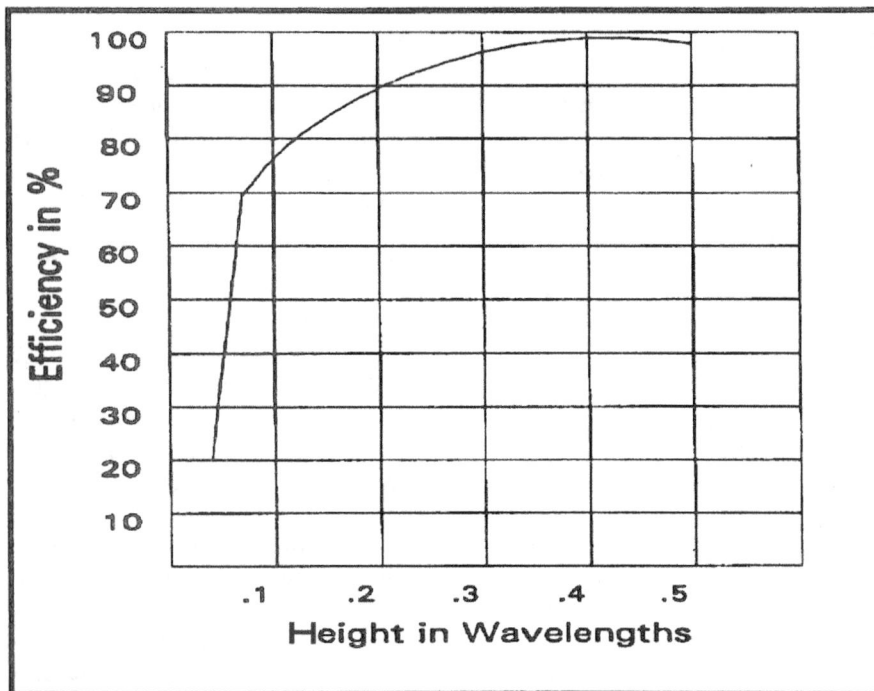

Figure 2 - Efficiency of a half-wave dipole over poor quality soil

the subsurface antenna will need to be only 50% of the length it would be in free space. While the antenna matching units that come with most standard radios will couple to an antenna that is electrically too long, efficiency and the critical effective radiated power will suffer if the antenna is too far from resonance.

In order to cut a dipole to the correct length, the buried antenna should be measured with a Return Loss Bridge, such as Antenna System Test Set TS-4351/PRC (NSN 6625-01-324-9273), and trimmed to a resonant length. Power Meter TS-4350/PRC (NSN 6625-01-323-6267) can also be used for this purpose. Both are components of Antenna Group OE-452/PRC (NSN 5985-01-279-7942) which is also known as the Special Operations Radio Antenna Kit (SORAK).

The actual procedure should be to construct an antenna and bury it in soil similar to that in the area of actual operations. Then test the antenna with these test sets and trim to the proper length for soil conditions and depth. This should done in a sanctuary area prior to deployment. It does take some time, but it's worth it.

Dipole construction: Due to their contact with the ground, the subsurface antenna must be well insulated. While antennas made of wire and inserted into a plastic pipe or hose using air as a dielectric have been shown to work,

construction takes considerable time, and the required materials add to the already heavy load of the deploying unit.

The best practical approach for radiating antenna wire has turned out to be standard coax cable (RG-8 or larger) with the outer cover and shield stripped off. This leaves the inner (center conductor) wire surrounded by a very good plastic dielectric for use as the radiating antenna elements.

Any standard dipole feed device such as the IL-4/GRA-4 NSN 5970-00-405-8223 will do to terminate the radiating elements. Connection to the feed point should be by a short coaxial cable kept at right angles to the radiator to avoid field distortion. The feed point and the ends of the radiating elements must be sealed to eliminate any path to ground for the antenna energy, due to water penetration. Similarly, the dielectric must be inspected to assure that there are no splits or cracks in it. Be very careful not to cut the dielectric when stripping the cover and shield off the coax cable to make the antenna elements.

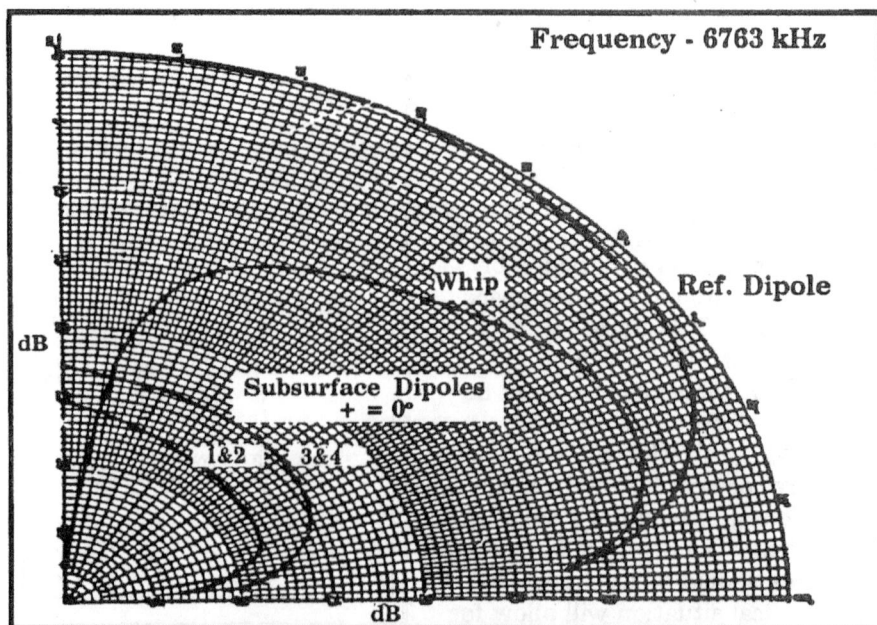

Figure 3 - Comparison of two element dipole arrays at different depths (1 and 3 feet) with standard single element dipole and whip. Note typical NVIS pattern but greatly reduced gain.

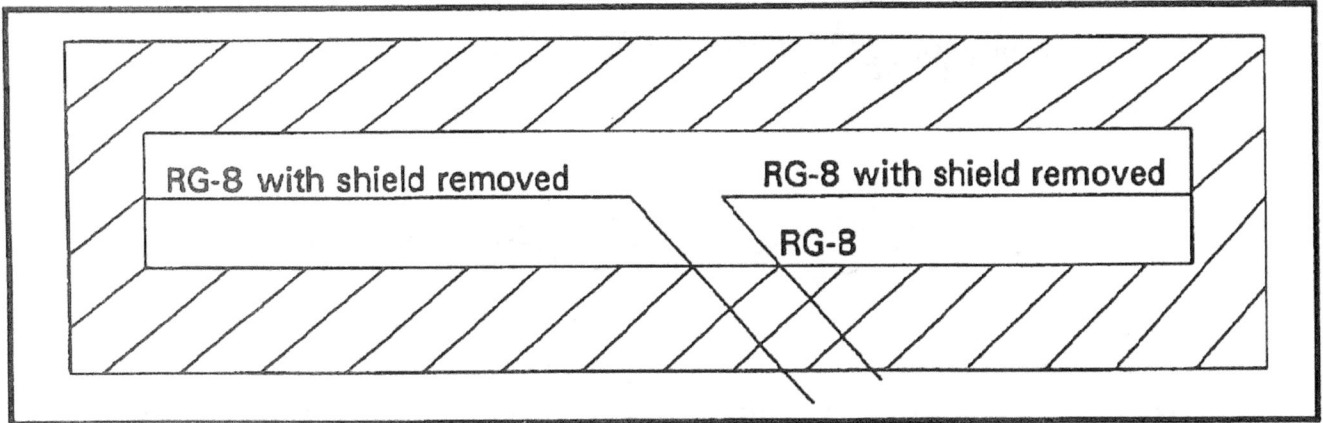

Figure 4 - Buried, insulated horizontal dipole center fed with ends and feed point sealed.

Dipole performance: The relative gain of a subsurface dipole two element array was measured by comparing field strength with a reference halfwave dipole one quarter wavelength off the ground (see fig 6).

In order to improve subsurface dipole performance, two parallel dipoles approximately 14 feet apart can be constructed and fed from a common feed point; this will increase the antenna gain (field strength) by 3db, which is equivalent to doubling the transmitter power (see fig 5).

Figure 6 is a plot of field strength for subsurface and

Units of US Air Force, the New Jersey Army National Guard and other organizations have employed subsurface antennas of the types described here for practical missions with great success. The points to remember when using these techniques are:

• The simplest antenna to use is a center fed half-wave dipole (doublet) or dipole array (to produce more gain) made from standard components.

• Since there is less antenna gain due to subsurface installation, it is important to pick the frequency carefully.

• It is important to use a well sealed wire surrounded by a good dielectric (center portion of a coax line, for example) to keep antenna current from running off into the ground.

• A dipole in the ground resonates at a much shorter physical length than one in air for the same frequency. Correct length should be measured with VSWR meter or Return Loss Bridge for maximum efficiency.

• Radiation pattern for subsurface dipoles is the same as a dipole a quarter-wave above ground, but radiated energy is greatly reduced. This pattern is ideal for NVIS

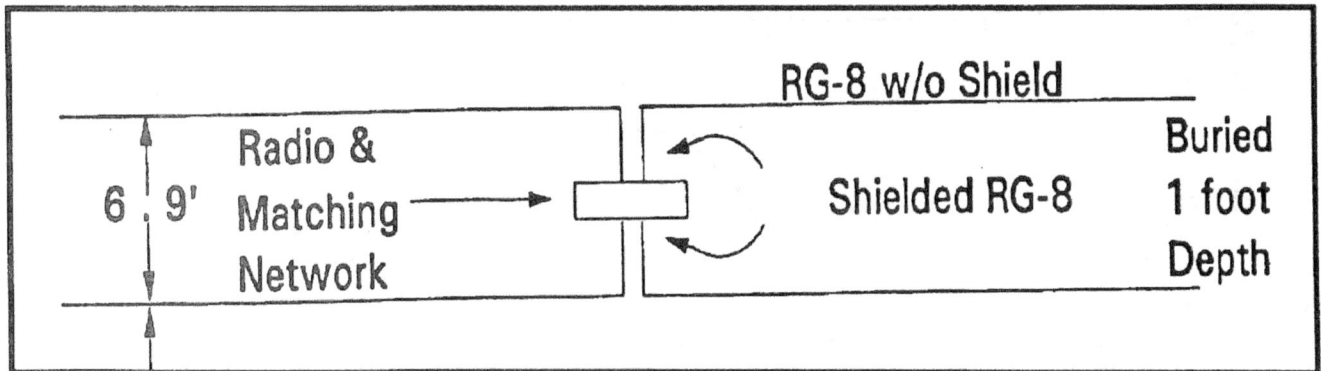

Figure 5 - 2 element buried antenna array. This configuration improves radiated signal by 3db.

reference dipoles. Burying the antenna results in loss of about 24db over an elevated antenna; however, sufficient signal still remains for communications (see side bar).

• The antenna should be buried only as deep as required for mission concealment (usually less than 6 inches) to minimize ground losses.

(0-400 mile) communication. For longer ranges, the dipole can be buried on a hill side sloping the path direction to a lower angle to get more gain in a desired direction.

COMPARISON OF REFERENCE AND SUBSURFACE DIPOLES WITH THEORETICAL "E" & "F" LAYER PROPAGATION

FIGURE 3-42

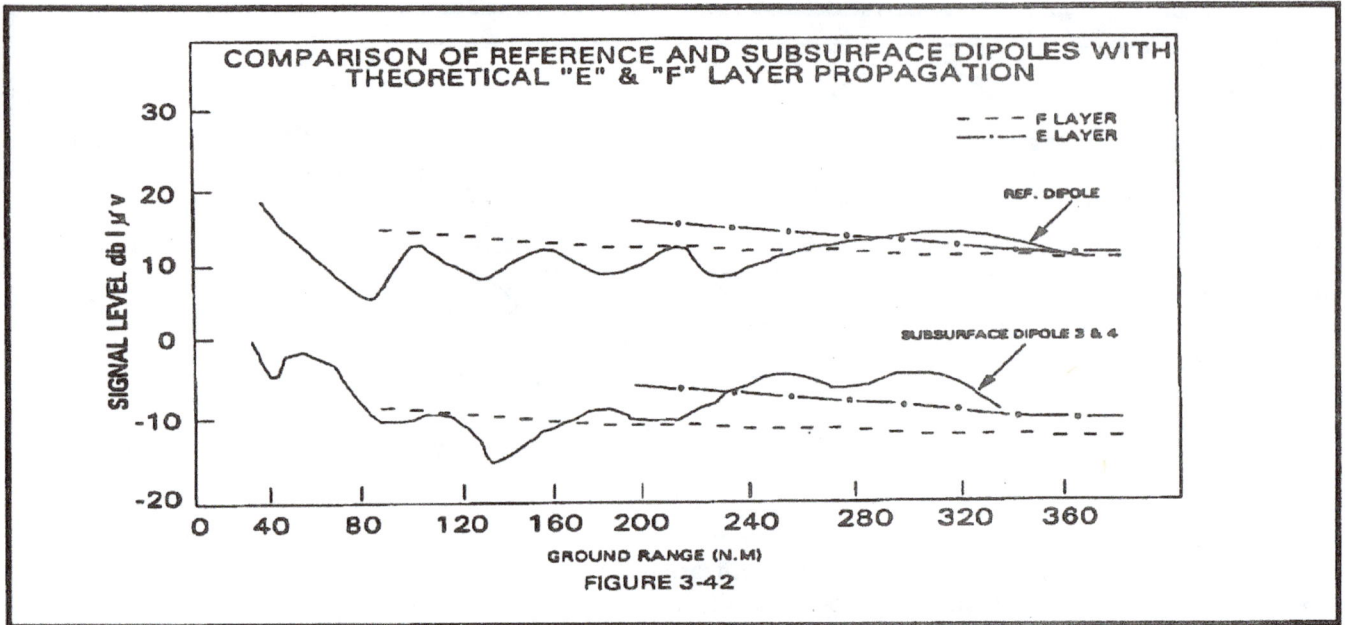

Figure 6 - Comparison of buried dipole array with reference dipole. Concealment under ground reduces signal strength by about 24 db.

• The buried dipole radiates some vertically-polarized groundwave energy off the ends. On short paths, this energy can be received by radio direction finders using whip antennas. Since this could be an enemy intercept station, it is desirable to orient the dipole ends away from possible intercept locations. Skywave communications are not affected by dipole orientations.

• Seal everything and install the antenna(s) in the driest spot available; water is the biggest problem with subsurface antennas. Be particularly careful to seal feed points and dipole ends.

Reasonable levels of reliable HF radio communications have been demonstrated with concealed dipoles of this type and standard manpack radios.

Use of these techniques will lower the probability of team detection and help assure that

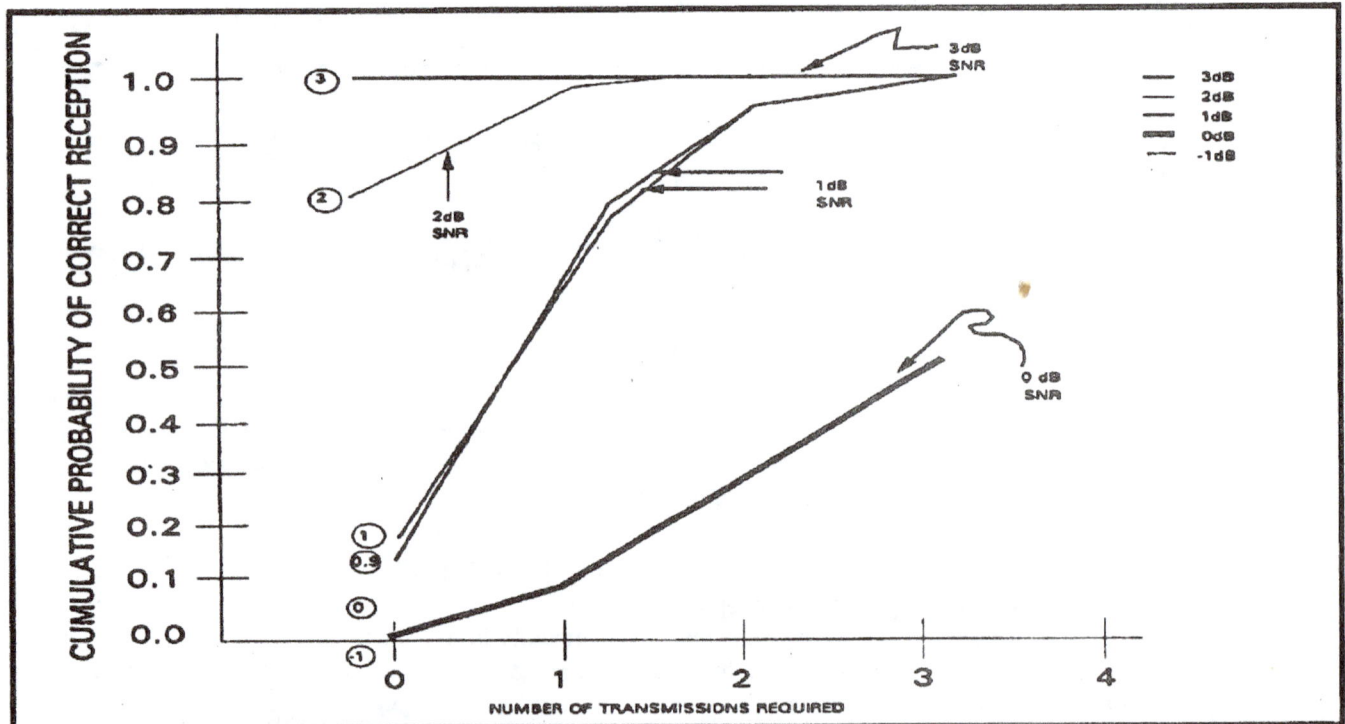

Figure 7 - System improvements resulting from use of AN/PSC-2 digital communications terminal. With 3db SNR, message was received every time. Requires a 10db SNR.

Sample calculations

Assuming a 400 watt radio at the base station with an efficient antenna and a 20 watt manpack radio at the LRSU/SOF out station using a subsurface two dipole antenna array of the type shown in Figure 5, we can calculate the path loss and required receive signal strength for the system to work:

Out station to base calculations

transmitter power--20 watts--+43db

free space path loss--110db (approximate)

subsurface antenna loss--20db (approximate)

total path loss (out station to base)--87db

receiver sensitivity--110db

signal to noise ratio required at receiver--10db above noise

communications margin (gains less losses 100--87 db)--13db (approximate)

This margin (13db) is often reduced by poor operating practices, such as weak radio batteries (3db additional loss), and poor frequency selection (up to 5db additional loss); however, sufficient power margin (5db) still does exist for successful communications even under these conditions.

This can be improved by up to 8db of processing gain by using data communication devices, such as the AN/PSC-2 Digital Communications Terminal (DCT) that use forward error correction (FEC) and packet retransmission (ARQ) techniques in terminal software (See Figure 7). Use of 50 watt manpack radios instead of 20 watts would also help by an additional 3+db.

System margin will increase when going from base to out station due to higher transmitter power gain (400 Watts=12db) and antenna gain (up to 20db over a subsurface antenna).

when we deploy a LRSU/SOF team, their HF communications will work and won't compromise unit security.

In view of the information presented here, it is clear that the Signal School and CECOM need to develop and analyze subsurface antenna requirements and communication techniques in order to provide the best combination of communications reliability and unit security possible for those engaged in a very dangerous mission.

References

The ARRL Antenna Compendium - Volume 1 Published by the American Radio Relay League, Newington, CT 06111, 1985.

United States Air Force-Rome Air Development Center Technical Report RADC-TR-69-221 "Measured Performance of Subsurface Dipoles" June 1969.

Mr. (LTC) Fiedler was commissioned in the Signal Corps upon graduation from the Pennsylvania Military College (Weidner University) in 1968. He is a graduate of the Signal Officers Basic Course, the Radio Microwave Systems Engineering Course, the Signal Officers Advanced Course, and the Command and General Staff College. He has served in Regular Army and National Guard signal, infantry, and armor units in both CONUS and Vietnam.

He holds degrees in both physic and engineering and an advanced degree in industrial management. He is employed as the Chief of the Fort Monmouth Field Office of the Project Manager All Source Analysis System (PM ASAS) and is the Assistant Project Director for the Integrated Meteorological System (IMETS). He is also assigned to the System Engineering Office of the Program Executive Office for Command and Control Systems (PEO CCS) with special responsibilities in the area of Battalion and Below Command and Control (B2C2) and Combat Command and Control Vehicles (C2V). Concurrently, he is the chief of the C-E Division of the NJ Army National Guard. A recognized expert in combat communications, he has published widely on communications topics.

11: ANTENNA PERFORMANCE FOR NEAR VERTICAL INCIDENCE SKYWAVE COMMUNICATIONS

The special requirements for antennas for NVIS systems have been described in the previous articles in this series [1,2] and in the work of other authors [4,5,6]. Now, let's take a detailed look at the theoretical and practical aspects of various antenna designs in NVIS service. This look relies on computer analysis using the NEC-2 antenna modeling methodology as well as practical experience. NEC-2 provides analytical data that is very difficult to generate any other way. Practical experience helps address the non-theoretical aspects of erecting and using antennas in the field.

Antenna Performance Estimators

Antenna performance is measured by gain in a particular direction. Gain is expressed in decibels (dB) relative to some standard. The most common standard is the isotropic radiator in free space. Isotropic means that it radiates equally in all directions (sort of like a bare light bulb) so its pattern plot from any direction appears to be a circle. Free space means just that — space in which nothing interacts with the antenna. Gain relative to an isotropic radiator is denoted as dBi.

When an antenna has gain with respect to an isotropic radiator it means that it concentrates its radiation in some particular direction (at the expense of other directions). The antenna does not make power, it merely concentrates (focuses) the power delivered to it by the transmitter through the feedline. For NVIS purposes, the antennas pattern should concentrate most of the radiation field vertically, at angles above 45 degrees, and should be round (omnidirectional) in azimuth. How well it concentrates its pattern is measured by beam width and beam extent [2].

While antennas don't "make" power they certainly can lose some of that provided to them. They do this by dissipating it in their internal resistance and by interactions with nearby objects, such as the earth. Sometimes these interactions are beneficial — for example in many cases reflections from earth improve radiation in certain directions. They can also be harmful — for example, RF currents can be dissipated in the earth's resistance. Most NVIS antennas have a radiation resistance much larger than their loss resistance [2] so efficiency is not much of a problem. Since NVIS antennas are generally mounted low to the ground, detrimental ground interactions can be a problem.

Practical considerations

Tactical communications usually involve base stations, field stations, and mobiles. Base stations are those at which there are the time and resources to erect optimal antennas. Field stations are those in which field expedient (but highly effective) antennas can be erected in a half hour or so. Mobiles, of course, are vehicles that must maintain the ability to be in motion, or to quickly initiate movement.

Base stations are the great opportunity in NVIS communication. LTC Fiedler has noted [4] the importance of exploiting opportunities for optimizing base station performance as a way of augmenting the more limited capabilities of field and mobile stations. This means operating at higher power and erecting more complex antenna arrays. It can also mean exploiting opportunities to reduce ambient RF noise reaching the base station's receiver. This is a big subject and will not be covered here.

Mobile stations are limited in antenna type to those which can be carried on the vehicle while it is in motion, or can be quickly deployed while it briefly pauses. LTC Fiedler has presented considerable research and practical information on this [4,7,8]. While this is an interesting subject it will not be directly covered here.

Field stations provide an opportunity to explore the factors that effect NVIS antenna performance in any situation while simultaneously exploring some of the performance vs.

convenience trade-offs involved. There is an incentive to use antennas that can be deployed quickly, can be depended upon to perform properly, can operate (or be adjusted to operate) over an adequate range of frequencies, and are sufficiently rugged to withstand prevailing conditions.

Most high frequency military field antennas will be built using some number of GRA-4 mast kits with GRA-50 antenna kits (or similar equipment). This allows a maximum height of about forty feet and includes sufficient wire to build an antenna resonant below 1.6 MHz. This equipment is adequate to construct a wide variety of antennas and consequently it will enable a unit to meet all NVIS mission requirements.

Rapid deployment

Deploying a wire antenna requires erecting the required number of masts, measuring and assembling the antenna, connecting the feedline, raising the antenna, and tuning the radio to the antenna. Some antennas also require that a grounding system (e.g., a counterpoise or radial system) be deployed. Most of the time and effort involves erecting the masts, particularly when guying is necessary. Consequently, when time is a factor there is an incentive to use as few masts as possible.

The California State Military Reserve considers a field team to consist of four persons. Such a team is expected to get a station on the air within thirty minutes of arrival on site. Most of that time is spent deploying an antenna. (Initially, the radio is powered by battery. Another half hour is often required to get a generator properly grounded, interconnected, and on-line.) We expect that under normal conditions three soldiers can erect a guyed GRA-4 mast in about ten minutes. A well-trained and motivated team can erect three masts, rig an antenna, and put a radio on the air within the allotted thirty minutes. Obviously, the deadline is a lot easier to meet when only one mast is involved. Antennas that require more masts need to do something worth the trouble. On the other hand, an antenna that is quick to deploy but does not perform well isn't of any value.

Antenna performance benchmarks and compromises

What works best? It is hard to improve on the performance of a resonant half-wave dipole mounted about 0.2 wavelengths above ground. No antenna performs significantly better. Its only liability is that it requires at least two masts, and for frequencies below 4 MHz Army doctrine requires three. If height is to be kept in the optimal 0.1 to 0.3 wavelength range the following table provides a guideline on how long the masts should be:

Table 1: Optimal Antenna Height		
Mast Height, ft.	Frequency (MHz) should be Above and Below	
20 ft.	4.7	14.0
30 ft.	3.1	9.4
40 ft.	2.3	7.0

Is there a way to get most of the dipole's performance without erecting two (or three) masts? An obvious compromise is a similar antenna that can be erected with a single mast: the Inverted V. As will be seen, its performance at the same mounting height is similar to the dipole and can be made nearly equal by increasing mast height while keeping the antenna as flat as possible. Installation convenience is gained by giving up a small amount of performance.

There are plenty of other alternatives and some of the better ones (see Figures 1a-1j) will be considered using the dipole as a benchmark.

Frequency agility

Practical around the clock NVIS operations require that we operate on more than one frequency. There is a substantial difference between day and night propagation so as a minimum two frequencies are essential. If automatic link establishment (ALE) is used to optimize propagation, a suite of several frequencies may be required.

The resonant dipole is a narrow-band device. Feedline losses, particularly when coaxial cable is used, can approach 3 dB as the frequency changes more than 15 percent. Yet the difference between day and night frequencies can be over 200 percent! At least two frequency changes are required during a 24-hour period. With wire antennas frequency changes are accomplished by lengthening or shortening the antenna wires. As long as the masts are set to accommodate the longest required antenna this is simply done. One shortcut is to string elements of two different lengths from the same feed point — one set forming a resonant dipole at the lower frequency with the other resonant at the higher. As long as frequency selection works as planned this simple technique provides near optimized operation on two frequencies without changing anything.

Antennas can be made broadband by loading which means adding a resistor somewhere in the design. Two examples are considered below — the T2FD (Figure 1f) and the End-Fed Terminated Wire (EFTW) (Figure 1j). Note that the inconvenience of physically modifying the antenna to change frequency is eliminated at the expense of a significant amount of gain (Figure 6).

Each antenna must be matched to the radio (by means of an internal or external antenna tuner). Losses in the feedline and matching system are not considered here. This is a significant simplification but it allows us to focus on the comparative performance of a large number of antenna designs.

Frequency agility can be addressed with designs in which careful selection of the antennas dimensions, type of feedline, and the means of matching it to the radio serve compromises that preserve both performance and convenience. This is a very interesting area but it is not the present subject.

Ruggedness

There are many tricks that can help get a station on the air quickly. For example, setup of an Inverted V using the GRA-4 with its tripod adapter can be managed by one soldier in a few minutes. This works well until the wind comes up. Another trick, which is regularly used when time is tight, is to use three guys per mast instead of four. Once again, this works well as long as the guys are well-positioned and in good soil. When the wind comes up on a dark and stormy night in which the ground has been softened by rain, the antenna can come down. This is embarrassing and inconvenient. It is usually easier to explain an extra few minutes spent getting on the air than it is to explain an extended forced outage at a critical moment. Always take every opportunity to enhance antenna ruggedness — even after a station is on the air.

Balanced and unbalanced antennas

Antennas can be broadly classified as electrically "balanced" and "unbalanced." All antennas have two sides — the current that promotes radiation must flow from somewhere to somewhere. Quarter-wave verticals, for example, find their second half in the earth. Inverted Vs find their second half in a counterpoise wire or in the earth. Antennas such as dipoles and Zepps have two obvious sides but not all antennas with two visible sides are balanced.

Consider a half-wave dipole. If one side is in the clear and the other is in trees, each side will interact differently with its environment.

An inclined dipole [11] has two obvious sides but one of them is much farther above ground than the other.

A sloper [12] uses the mast for one side and an inclined wire for the other. The feed point is at the top of the mast, hence there is a marked difference in geometry between the two sides.

Each of these situations results in different currents in each side and consequently unbalanced currents on the feedline. This situation promotes feedline radiation, and hence, pattern distortion. Pattern distortion in this case means that some radiation we need for NVIS effect is going in some less useful direction, e.g., horizontally.

For present purposes, the following antennas will be considered balanced: Resonant half-wave dipole, AS-2259 Inverted V, T2FDV, and Zepp.

The following antennas have two obvious sides but lack symmetry. They don't require a counterpoise or ground radial system but may suffer pattern distortion from feedline radiation: Inclined dipole; Sloper.

The following antennas are inherently unbalanced and require some form of counterpoise or ground radial system if pattern distortion is to be minimized and efficiency maintained: Inverted L, Sloping wire, End-Fed Terminated Wire.

Figure 1. Diagrams of some NVIS antennas

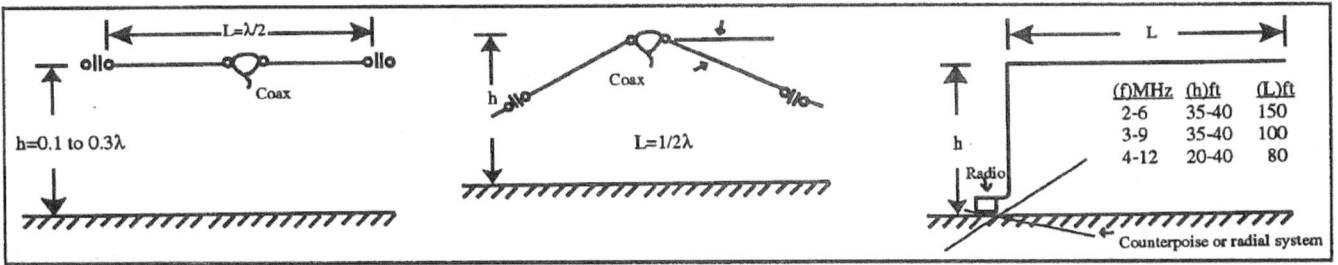

a. Half-wave Dipole   **b. Inverted V**   **c. Inverted L**

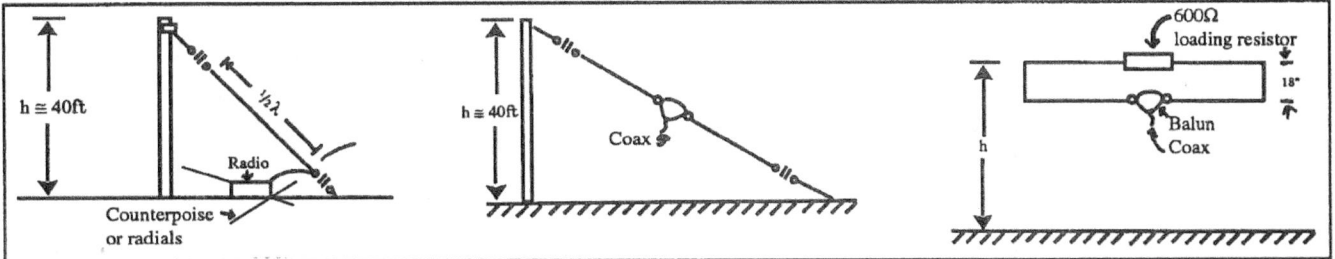

d. Sloping Wire   **e. Inclined Dipole**   **f. T2FD**

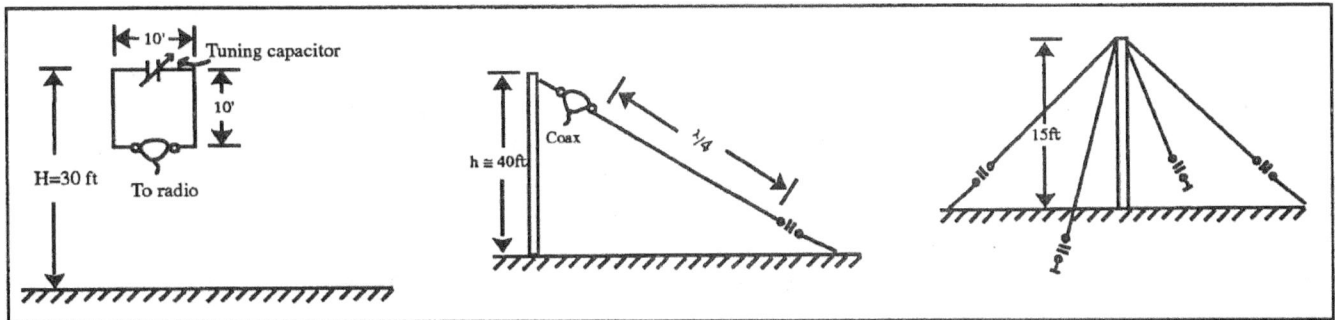

g. Small Loop   **h. Sloper**   **i. AS-2259**

(See [9] for general details). Scale distorted for clarity.

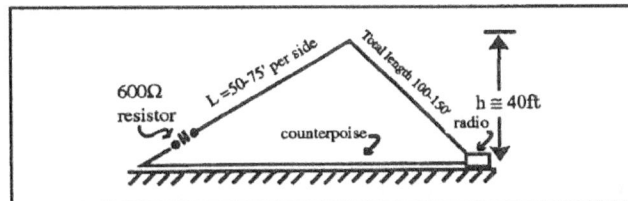

j. End-fed Terminated Wire

Starting with the half-wave dipole...

There is widespread agreement that a half-wave dipole (Figure 1a) mounted 0.1 to 0.3 wavelengths above ground is excellent for NVIS communication. Consequently, that antenna provides an excellent reference against which others can be compared. The key performance parameters are vertical gain, beam width, and beam extent. The key installation-related parameters are ground conductivity and mounting height.

A full-fledged dipole is fairly time consuming to erect. It requires at least two masts and below 4 MHz, possibly a third in the middle. While it seems as though the middle mast could be eliminated the long antennas frequently required for nighttime NVIS work suffer this poorly. The droop in the center forms a V antenna which effectively lowers the height. Also, without the center mast the strain on the two end masts is substantially greater thus making the antenna more susceptible to wind, poor guying conditions, and casual neglect.

Ground conductivity and its effect on vertical gain

Ground conductivity depends on the earth under the antenna. In order to avoid a lengthy discussion of ground conductivity and permittivity we will characterize ground as follows [9]: Figure 2 shows the vertical pattern plot of a half-wave dipole mounted at 0.2 wavelengths above each type of ground shown in Table 1. Note that ground properties produce a change of about 3 dB from the best to the worst. Three dB is equivalent to changing power by a factor of two. A 100-watt radio operating over very poor earth would produce about the same signal as a 50-watt radio operating over sea water.

Height above ground

There are two components to antenna height — the part you see and the part you don't see. The part you see is that represented by the length of the mast holding the wire above the earth. The part you don't see is the portion below the apparent surface through which radio waves travel before reflecting. For a very conductive ground, such as sea water, radio waves reflect from the surface. For a poor ground, such as fresh water, radio waves penetrate many feet. An antenna lying on the surface of rocky soil, for example, might have an effective "height" of 40 feet or more. This depth of penetration depends on the frequency as well as the ground itself. Ground is lossy and so it isn't as though "a mirror moves up and down." All computer-generated pattern plots presented herein consider this effect.

Figure 3 shows the vertical gain pattern of a half-wave dipole at various heights (in wavelengths) above average ground. Figure 3 also shows the vertical gain as a function of mounting height. Note that the vertical gain peaks at about 0.2 wavelengths — above that height the antenna begins to develop its characteristic "bat wings" which concentrate a higher percentage of the radiation at lower angles. At lower heights, ground interaction be-

gins to consume the signal. Performance between 0.1 and 0.3 wavelengths is fairly constant — it varies about 1.5 dB. As the antenna is lowered further the gain drops off rapidly. When the antenna is lying just above the ground gain is reduced by about 15 dB from that at 0.2 wavelengths. Put another way, if we compare our 100-watt radio with an antenna lying on the ground against the same radio connected to an antenna at 0.2 wavelengths, we would discover that our signal was about the same as if the transmitter power were reduced to about 3 watts. It is important to note, how-

Effect of Ground on Vertical Gain of a Dipole Antenna Mounted at 0.2 Wavelengths				
Ground Type	for Example	Legend	Conductivity Siemens/mtr	Permitivity
Very poor	Cities, industrial areas	VP	0.001	5
Poor	Rocky soil, steep hills, mtns.	P	0.002	13
Average	Pastoral, heavy clay, med. Forestation	A	0.005	13
Fresh water	Fresh water	FW	0.001	80
Very good	Pastoral, low hills, rich soil	VG	0.0303	20
Salt water	Salt water	SW	5	81

Table 2: Characteristics of Earth

ever, that the pattern shape remains the same.

For mounting heights of 0.2 wavelengths and below, the beamwidth and beam extent are nearly constant at about 135 degrees and 160 degrees, respectively. This beamwidth is more than ample to provide signal strengths within 3 dB over an area with a radius of 400 km or so. The beam extent is so broad that this antenna does not provide much attenuation for signals arriving from far away (at low angles).

Study these figures. They will help you develop a "feel" for the performance you can expect in each situation you will encounter.

The Inverted V

An Inverted V (Figure 1b) can be erected using a single mast in its center. This is much simpler and quicker than erecting a dipole.

It will perform almost exactly like a dipole mounted at a slightly lower height. Figure 4 shows the elevation pattern with the center at 40 feet and with various amounts of droop. Clearly, the flatter the better.

If you think about it, this is logical. It is RF current in horizontal wire sections that produces NVIS effect. The sharper the V, the less current there will be in the horizontal plane. Further, the fact that placing a V in the an-

Dipole Over Various Grounds
All At 0.2 λ Above Ground
11-04-1994   15:55:11
Freq = 5 MHz

EZNEC 0.02

Sea Water
Very Good Earth
Fresh Water
Good Earth
Poor Earth
Very Poor Earth

0 dB
-10
-20
-30
0 deg.

Outer Ring = 8.066 dBi
Max. Gain = 8.066 dBi

Elevation Plot
Azimuth Angle = 0.0 Deg.

Vertical Gain of Dipole at 0.2 Wavelengths Over
Various Types of Ground

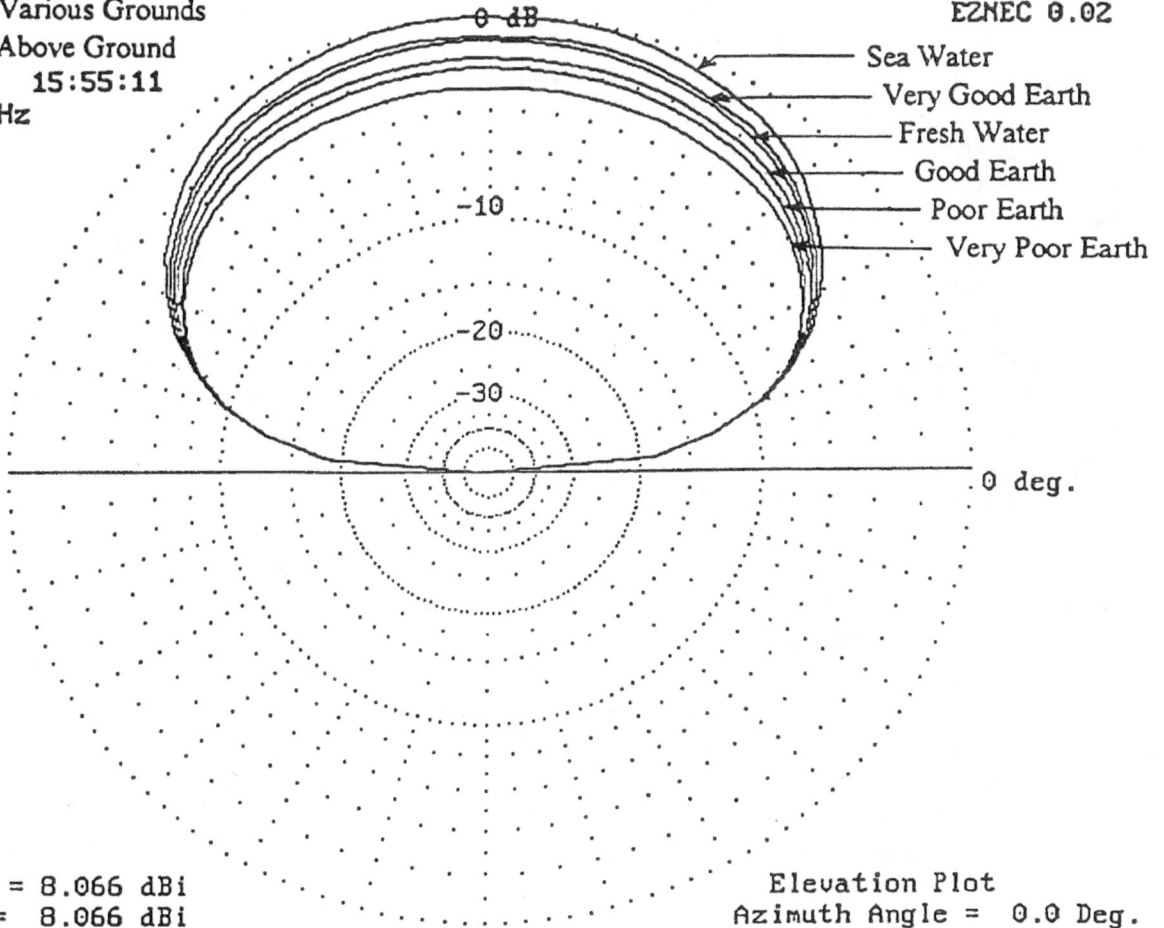

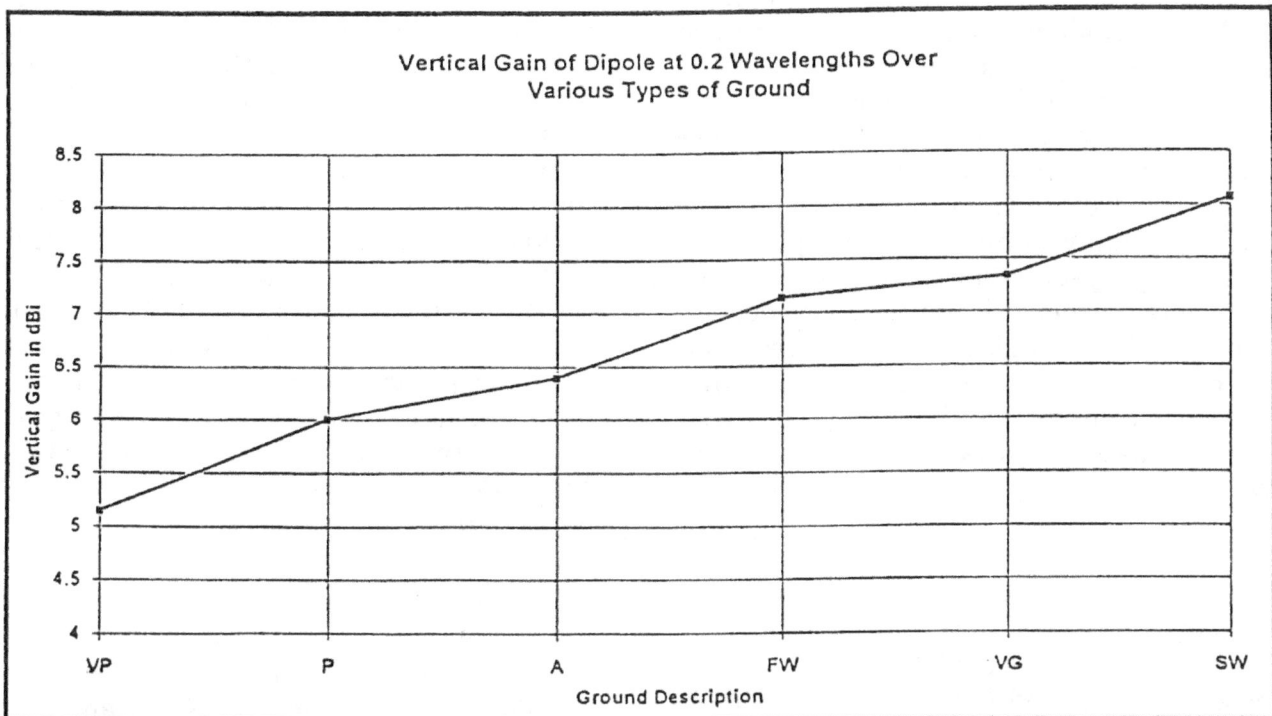

Figure 2. Half-wave dipole at 0.2 wavelengths over various grounds. While not much can be done to change the characteristics of the ground that is encountered, it is comforting to see that from the best to the worst the impact on radiated signal is only about 3 dB.

tenna effectively lowers it is also logical. The current that produces radiation is highest at the center of the antenna and it decreases sinusoidally along the wire arms, finally becoming zero at the end. When the antenna is not flat it is obvious that some of this radiation-producing current is flowing at lower height so we would expect the effect to be that of lowering the antenna. If you work through the math you will discover that a point on an element that is one third of the distance from the center to the end is useful for estimating the effective height of the entire antenna. A similar analysis can be made for the V antenna in which the center is lower than the ends.

The Center-fed Extended Double Zepp

If RF current in wire produces radiation then more wire should be better. This is true up to a point. As the wire becomes longer the pattern tends to become directional in azimuth as well as elevation.

An extended double Zepp antenna is a dipole in which each side is $5/8$(0.625) wavelength long. Obviously this antenna is over two and a half times the size of a standard half-wave dipole. Figure 5 shows the vertical gain (over average ground) compared with a half-wave dipole. Note that the Zepp produces over three dB of gain, thus converting our 100-watt transmitter to an effective power of 200 watts! The bad news is also shown in Figure 5 in which it is apparent that we only get all this gain along the axis, or perpendicular to the axis of the antenna. Note there are deep nulls at 45 degrees to the axis. Stations located in the direction of these nulls would find our signal down by over 10 dB when compared with a half-wave dipole. This serves as an important reminder that long antennas are usually directional in azimuth.

T2FD

While opinions vary, T2FD (Figure 1f) seems to stand for Tactical Terminated Folded Dipole or Terminated Two-wire Folded Dipole. (See [6], pp. 50, 51, 173,181) This antenna has also been called the squashed rhombic. A commercial version is manufactured by B&W. Its primary virtue is that it has a very high SWR bandwidth. This is achieved by loading the antenna in the center of the top wire with a 600-ohm non-inductive resistor. As with all loaded antennas, gain suffers significantly. Figure 6 shows the relative performance of the T2FD.

AS-2259/GR

This antenna (Figure 1I) was designed for use with the AN/PRC-47 and other manpack radios. [11]. It produces NVIS effect and can be tuned over a broad frequency range. It consists of a 15-foot mast which doubles as the coaxial feedline and four radiating elements of two different lengths which double as guys. Its performance depends on frequency but at 5 MHz you can gauge the performance relative to some other NVIS antennas in Figure 6. Its performance relative to a resonant dipole degrades as frequency is decreased. This antenna is tricky to tune and is notorious for problems in the mast/feedline assembly. In most cases a standard dipole is as easy to install and performs significantly better.

Throw it on the ground

Eyring of Provo, Utah is marketing an antenna they call the ELPA for "Extremely Low Profile Antenna." Some of these antennas have been purchased by U.S. and Canadian military units. Their design is very broadband and very large. See Figure 7. It is interesting to compare the NVIS performance this antenna (using NEC-2) with a dipole mounted at a similar height (three inches in this example). In this case the Eyring is set up in the 75 x 75 configuration they recommend for NVIS applications in this frequency range. Both antennas have omnidirectional azimuth patterns. Figure 7 compares their vertical gains at 5 MHz. Although larger and more difficult to install the Eyring exhibits a significant 4.5 dB advantage over the dipole.

Unbalanced Antennas

The usual unbalanced NVIS antenna is made by attaching one end of a wire to the radio and running the other end up and off in the horizontal direction for some convenient distance.

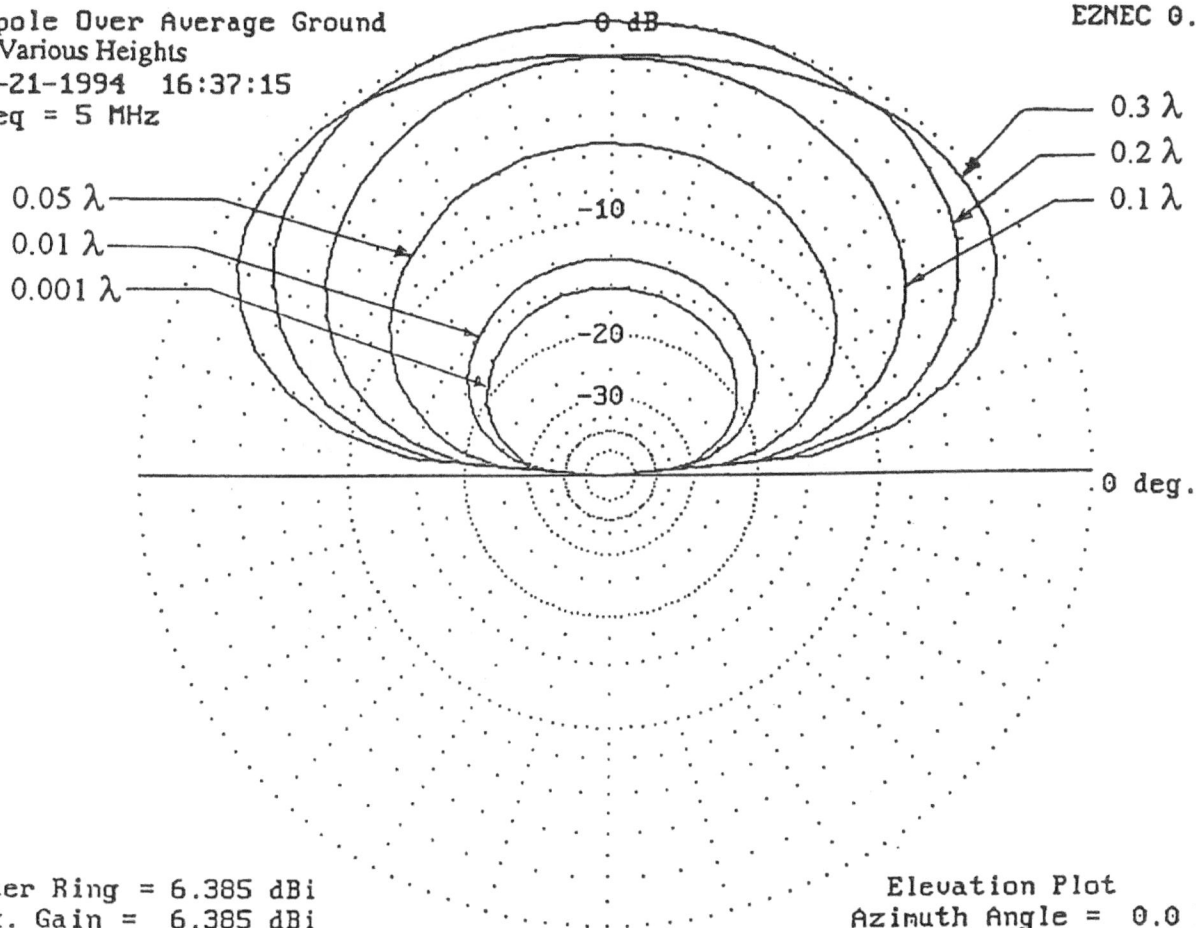

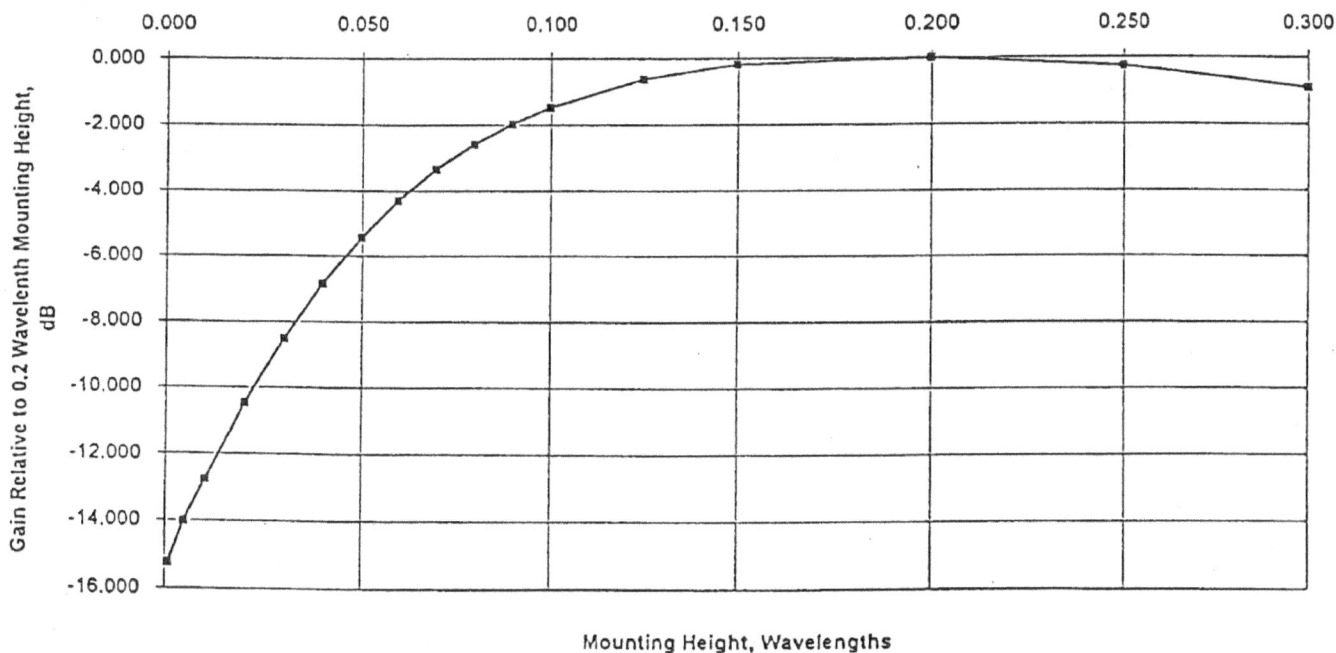

Figure 3. Pattern of a half-wave dipole at various heights (in wavelengths) above ground. Optimal performance for NVIS purposes is obtained at 0.2 wavelengths. Proper antenna height is essential for good NVIS performance. For an excellent treatment of the (non-NVIS) performance of a higher-mounted dipole, see [9], pg. 3-8, 9.

Inverted V

09-26-1995   15:29:46
Freq = 5 MHz

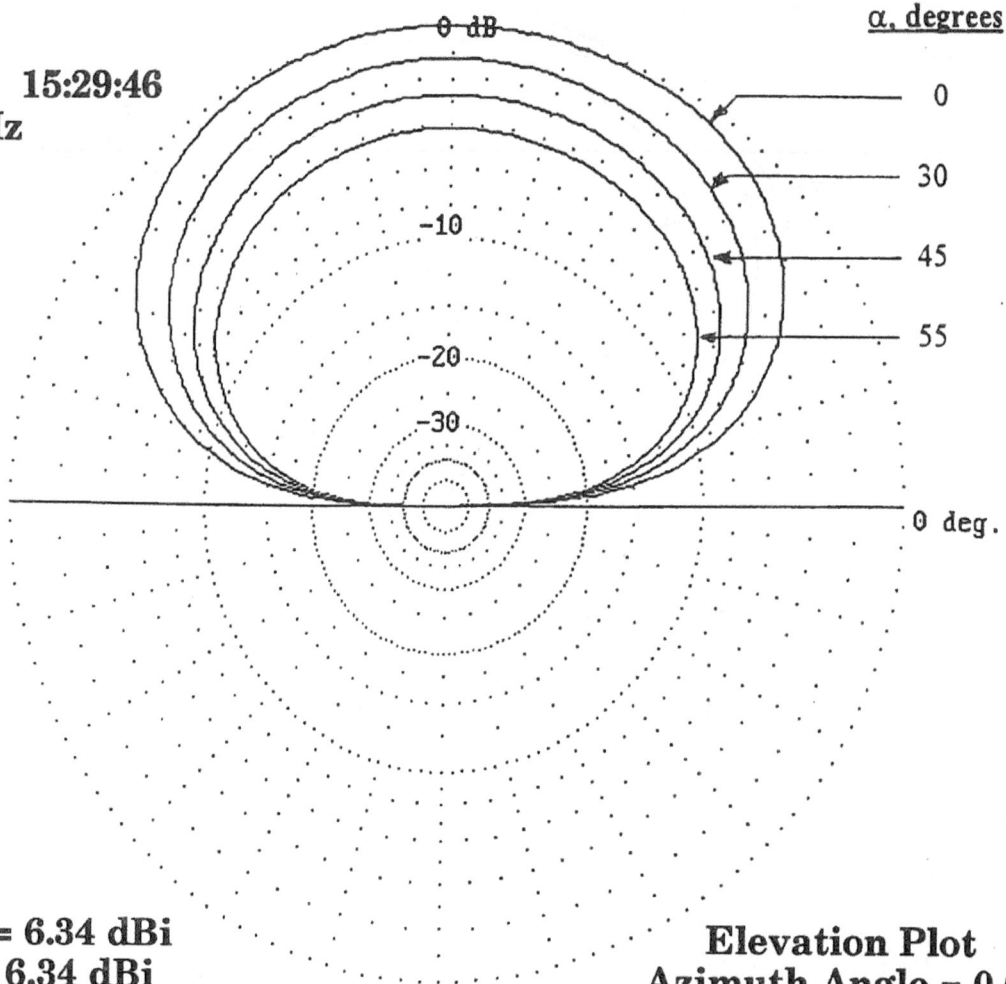

0 dB

α, degrees
— 0
— 30
— 45
— 55

−10

−20

−30

.0 deg.

Outer Ring = 6.34 dBi
Max. Gain = 6.34 dBi

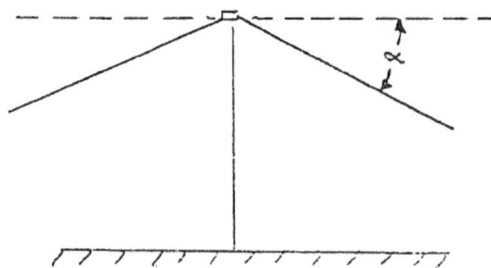

Elevation Plot
Azimuth Angle = 0.0 Deg.

Figure 4. A 5 MHz Inverted V with the center at 40 feet but with various amounts of droop. Clearly, keeping the "V" shallow enhances performance. A 45-degree droop costs 3 dB.

If the antenna slopes uniformly up from the radio to its highest point we call it an inclined wire or random wire.

If it rises straight up to some height and then runs horizontally we call it an Inverted L. (Actually, the "classical" inverted L has a total length of about 0.5 wavelength. Such an antenna does not perform well for NVIS work —

too much of the current is in the vertical section.)

As always, the performance is determined by where the RF current occurs. Remember, for NVIS work we want most of the current in a horizontal section that is between 0.1 and 0.3 wavelengths above ground.

In Figure 6, it is simple to compare an

Inverted L with several NVIS antennas. In this case the Inverted L rises up 40 feet and then runs 100 feet horizontally. While the vertical gain is about 4 dB less than the dipole, the on-axis beam width is a narrow 52 degrees and the beam extent is less than 90 degrees. While the vertical gain is lower, this Inverted L provides significant attenuation for distant signals that would be heard as interference if the dipole were used. Inverted Ls are directional and they require a counterpoise or radial system. For this analysis a four-radial system, with each radial 40 feet long, was used. Unfortunately the azimuth pattern (see Figure 8) shows significantly less gain on-axis than across-axis. Orientation is very important.

Field experience indicates this antenna, when properly erected, performs similarly to an Inverted V (which Figure 6 also bears out). The installation is actually somewhat more difficult than required for the Inverted V because of the counterpoise or radial system.

End-Fed Terminated Wire (EFTW)

The end-fed terminated wire (Figure 1j, see [6], pp. 117, 163) is another loaded antenna. As with all loaded antennas, some of the

transmitter's power produces nothing but heat in the loading resistor. As Figure 6 shows, it is not an outstanding performer for NVIS work. It is actually somewhat better for long distance work as it produces significant radiation at low angles.

The Sloper and Inclined Dipole

The sloper (Figure 1h) consists of a wire sloping from the ground to the top of a mast. It is fed at the top of the mast in such a way that the mast is connected to the shield of the coax and the sloping wire is connected to the center conductor. The pattern from a sloper made using a 40-foot mast and a half-wavelength sloping wire is shown in Figure 6. A similar approach involves replacing the sloping wire with a center-fed dipole and eliminating the connection to the mast. This antenna is called an inclined dipole (Figure 1e). Its performance is also shown in Figure 6. (For details, see [6], pp. 148, 149 and [9], pp. 4-15 to 4-18.)

Small Loops

A compact antenna (Figure 1g) can be made by arranging conductor in a loop. (see [9],

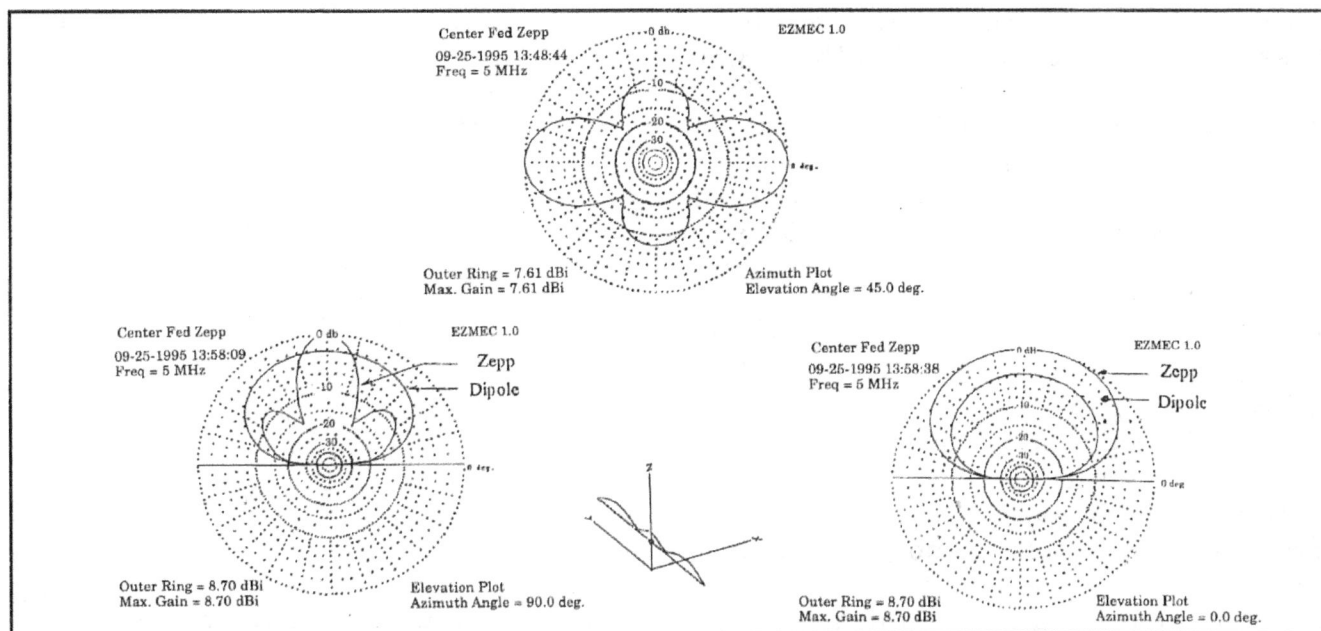

Figure 5. The dipole can be out-performed by the center-fed Zepp, but only in certain directions. In NVIS work, where omnidirectional coverage is usually the goal, it is important to keep in mind that as length increases, the antenna becomes directional. This phenomenon effects all antennas that are long — in terms of wavelengths.

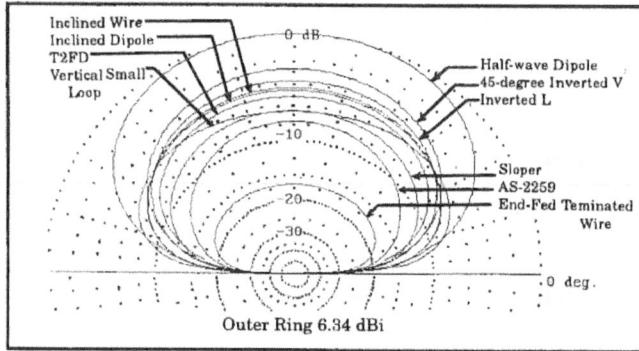

Figure 6. Performance of ten NVIS antennas at 5 MHz is shown. Note that a half-wave dipole has nearly a 3 dB advantage on any of them.

inclined dipole (Figure 1e.) Its performance is also shown in Figure 6. (For details, see [6], pp. 148, 149 and [9], pp 4-15 to 4-18.)

Small Loops

A compact antenna (Figure 1g) can be made by arranging conductor in a loop. (see [9], pp. 5-2 and 5-11 to 5-17) The loop is fed in the middle of one side and tuned by means of a capacitor installed in the middle of the opposite side. The radiation resistance of this arrangement is very small, less than one ohm. Efficiency requires that all ohmic losses, in the connections and in the tuning capacitor, be minimized. Resistance of the loop itself must also be kept very small, which usually requires that the loop be made from large diameter copper pipe. For present purposes a square loop ten feet on a side was modeled. Optimum NVIS effect was produced with the loop arranged in the vertical plane and mounted with the top at thirty feet. Its performance can be evaluated by reference to Figure 6.

Small loops exhibit very high Q hence tuning is very critical — they must be carefully adjusted for each operating frequency. Broadband operation without retuning is out of the question. Tuning involves precisely manipulating the loop's variable capacitor, which is difficult to do quickly. Since current in the loop is very high and the voltage that appears across the capacitor can be very large, automatic tuners are not a good choice. Large metal objects, such as ar-

mored vehicles, moving near a loop will cause it to require retuning.

Conclusion

There are many antennas suitable for field NVIS applications. It is hard to improve on the resonant half-wave dipole and Inverted V. Practical considerations, in addition to performance, drive antenna selection in military applications. Rapid deployment and the ability to withstand the conditions of the day are important. Within broad limits most environmental effects, such as ground conductivity, have less effect than se-

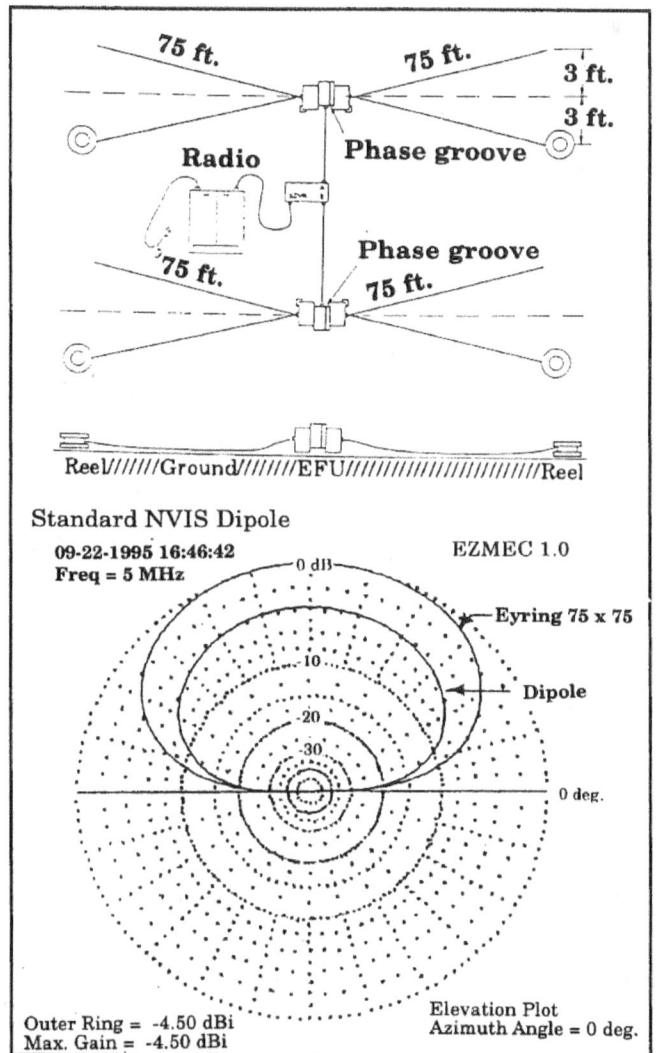

Figure 7. When properly configured in the Eyring - recommended "75 x 75" layout, their ELPA has nearly a 4.5 dB advantage over the simpler-to-erect half-wave dipole. Both antennas are omnidirectional azimuth.

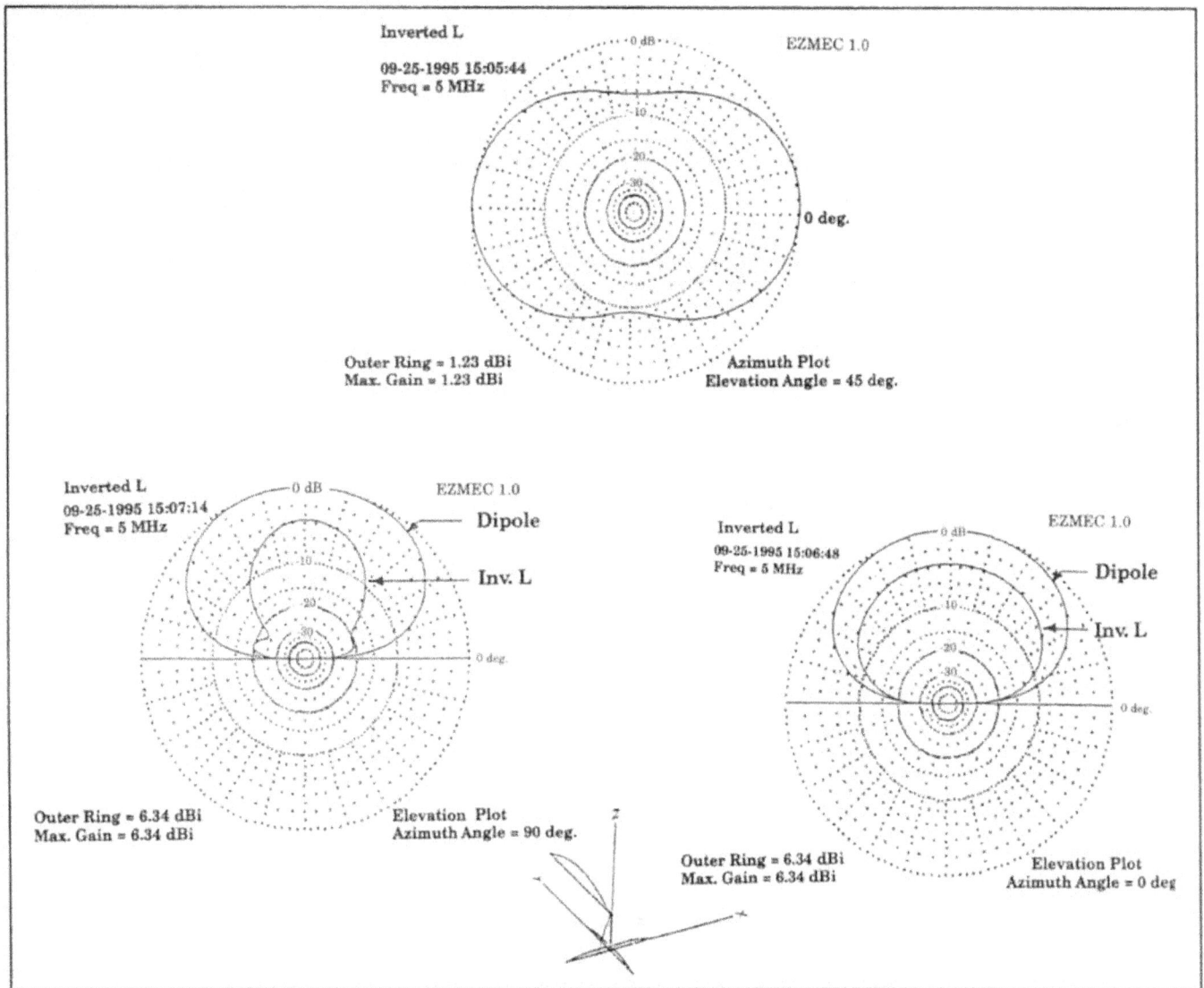

Figure 8. The Inverted L has directional aspects to its pattern. Note that in certain directions that beamwidth and beam extent are significantly narrower than a dipole, although vertical gain is lower.

lection of the proper antenna type.

In most situations, a unit that pays proper attention to basics and does a professional job of installation will find they are the CE heroes of the operation.

Footnotes:

1. Farmer, Edward J.; NVIS Propagation at Low Solar Flux Indices; **Army Communicator** Magazine; Spring 1994; Vol. 19, No. 1; Ft. Gordon, GA.

2. Farmer, Edward J.; NVIS Antenna Fundamentals; **Army Communicator** Magazine; Spring 1994; Vol. 19 No. 3; Ft. Gordon, GA.

3. Farmer, Edward J.; A Look at NVIS Techniques; **QST** magazine; American Radio Relay League; January 1995; Volume 79, Number 1.

4. Fiedler, LTC David M.; Optimizing Low Power High Frequency Radio Performance for Tactical Operations; **Army Communicator** magazine; Spring 1989; Ft. Gordon, GA.

5. Fiedler, LTC David M.; HF Radio Communi-

cations and High Angle Antenna Techniques; **Army Communicator** magazine; Summer, 1991; Ft. Gordon, GA.

6. Christinsin, Alan S.; **Tactical HF Radio Command and Control — An Anthology**; ASC & Associates, Ltd.; Belleville, IL; Copyright 1993.

7. Fiedler, LTC David M.; Mobile NVIS: **The New Jersey Army National Guard Approach**; **Army Communicator** magazine; Fall, 1987; Ft. Gordon, GA.

8. Fiedler, LTC David M.; Marine Tests Prove Fiedler's NVIS Conclusions; **Army Communicator** magazine; Fall, 1989; Ft. Gordon, GA.

9. Hall, (Editor); **The ARRL Antenna Book**; 15th Edition; Chapter 3; The American Radio Relay League; Newington, CT; Copyright 1988.

10. **Operations Manual for the 302A Eyring Low Profile Antenna**; Eyring Document Number 300-0086; Eyring, Inc.; Provo, UT; Copyright 1990.

11. TM-11-5985-379-14&P; The AS-2259/GR Antenna; U.S. Army.

12: U.S. ARMY FM 24-18: TACTICAL SINGLE-CHANNEL RADIO COMMUNICATIONS TECHNIQUES

ANTENNAS

Section I. Requirement and Function

3-1. Necessity

   All radios, whether transmitting or receiving, require some sort of antenna. Single-channel radios normally send and receive radio signals on one antenna. This is called one-way-reversible (OWR) or simplex operation. During duplex (DX) operation two antennas are used, one for transmitting and the other for receiving. In either case, the transmitter generates a radio signal. A transmission line delivers the signal from the transmitter to the antenna. The transmitting antenna sends the radio signal into space toward the receiving antenna. The receiving antenna intercepts the signal and sends it through a transmission line to the receiver. The receiver processes the radio signal so that it can either be heard or used to operate a recording device such as a teletypewriter (fig 3-1).

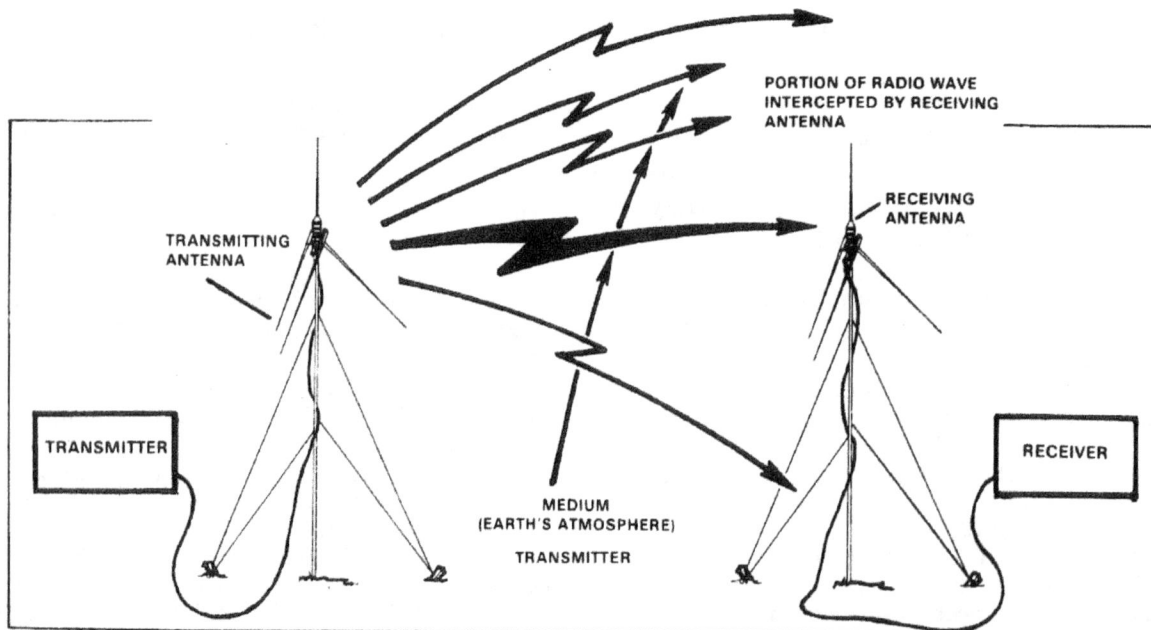

Figure 3-1. Simple radio communications network.

3-2. Function

The function of an antenna depends on whether it is transmitting or receiving. A transmitting antenna transforms the output RF energy produced by a radio transmitter (RF output power) into an electromagnetic field that is radiated through space. In other words, the transmitting antenna converts energy from one form to another form. The receiving antenna reverses this process. It transforms the electromagnetic field into RF energy which is delivered to a radio receiver.

3-3. Gain

The gain of an antenna depends mainly on its design. Transmitting antennas are designed for high efficiency in radiating energy, and receiving antennas are designed for high efficiency in picking up energy. On many radio circuits, transmission is required between a transmitter and only one receiving station. In this case, energy may be radiated in one direction because it is useful only in that direction. Directional receiving antennas increase the energy pickup or gain in the favored direction, and reduce the reception of unwanted noise and signals from other directions. The general requirements for transmitting and receiving antennas are that they have small energy losses and that they be efficient as radiators and receptors.

Section II. Characteristics

3-4. Electromagnetic Radiation

Radiation Fields.

When RF power is delivered to an antenna, two fields are set up: one is an induction field, which is associated with the stored energy; the other is a radiation field. At the antenna, the intensities of these fields are large and are proportional to the amount of RF power delivered to the antenna. At a short distance from the antenna and beyond, only the radiation field remains. This radiation field is composed of an electric component and a magnetic component (fig 3-2).

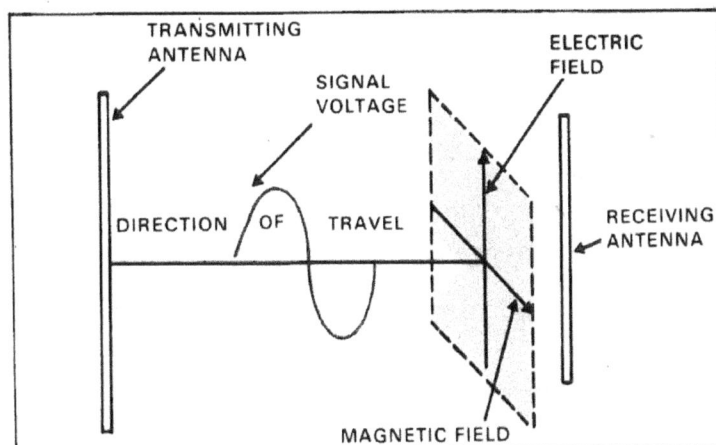

Figure 3-2. Components of electromagnetic waves.

The electric and magnetic fields (components) radiated from an antenna form the electromagnetic field. The electromagnetic field is responsible for the transmission and reception of electromagnetic energy through free space. A radio wave is a moving electromagnetic field that has velocity in the direction of travel, and with components of electric intensity and magnetic intensity arranged at right angles to each other.

Radiation Patterns.

The radio signals radiated by an antenna form an electromagnetic field having a definite pattern, depending on the type of antenna used. This radiation pattern is used to show the directional characteristics of an antenna. A vertical antenna theoretically radiates energy equally in all directions (omnidirectional); a horizontal antenna is mainly bidirectional. There are also unidirectional antennas. These antennas theoretically radiate energy in one direction. In practice, however, the patterns usually are distorted by nearby obstructions or terrain features.

The full- or solid-radiation pattern is represented as a three-dimensional figure that looks somewhat like a doughnut with a transmitting antenna in the center (fig 3-3). The upper pattern in the figure is that of a quarter-wave vertical antenna; the center pattern is that of a half-wave horizontal antenna, located one-half wavelength above the ground. The bottom pattern is that of a vertical half-rhombic antenna.

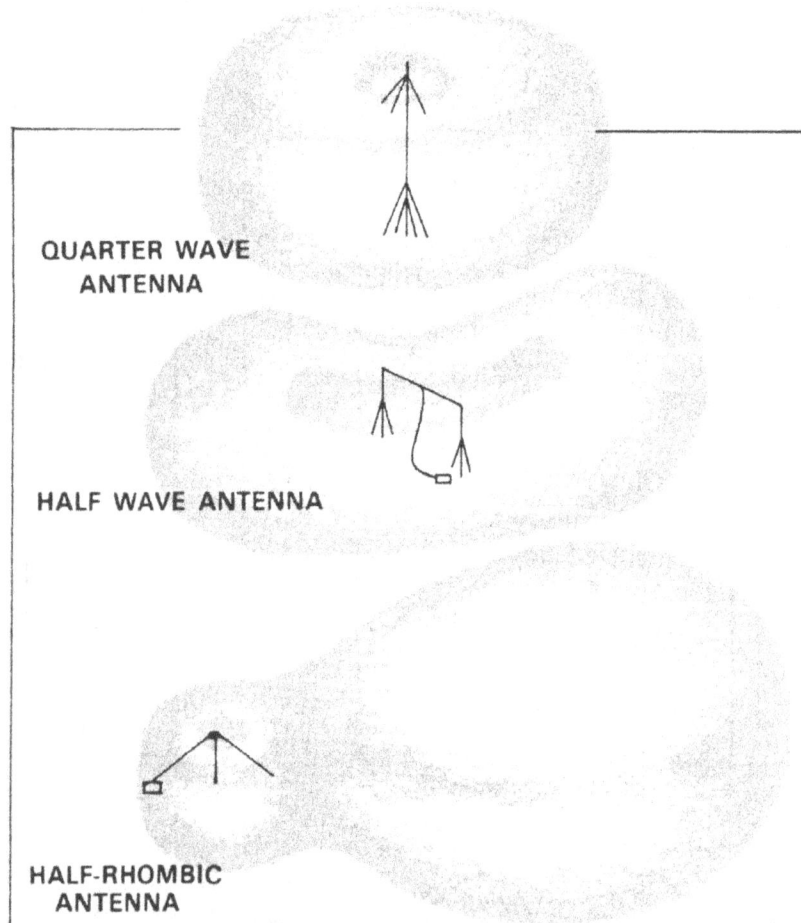

QUARTER WAVE
ANTENNA

HALF WAVE ANTENNA

HALF-RHOMBIC
ANTENNA

Figure 3-3. Solid radiation patterns from quarter-wave, half-wave, and vertical half-rhombic antennas.

3-5. Polarization

The polarization of a radiated wave is determined by the direction of the lines of force making up the electric field. If the lines of electric force are at right angles to the surface of the Earth, the wave is said to be vertically polarized (fig 3-4). If the lines of electric force are parallel to the surface of the Earth, the wave is said to be horizontally polarized (fig 3-5). When a single-wire antenna is used to extract (receive) energy from a passing radio wave, maximum pickup results if the antenna is oriented so that it lies in the same direction as the electric field component. Thus, a vertical antenna is used for efficient reception of vertically polarized waves and a horizontal antenna is used for the reception of horizontally polarized waves. In some cases, the field rotates as the waves travel through space. Under these conditions, both horizontal and vertical components of the field exist and the wave is said to have elliptical polarization.

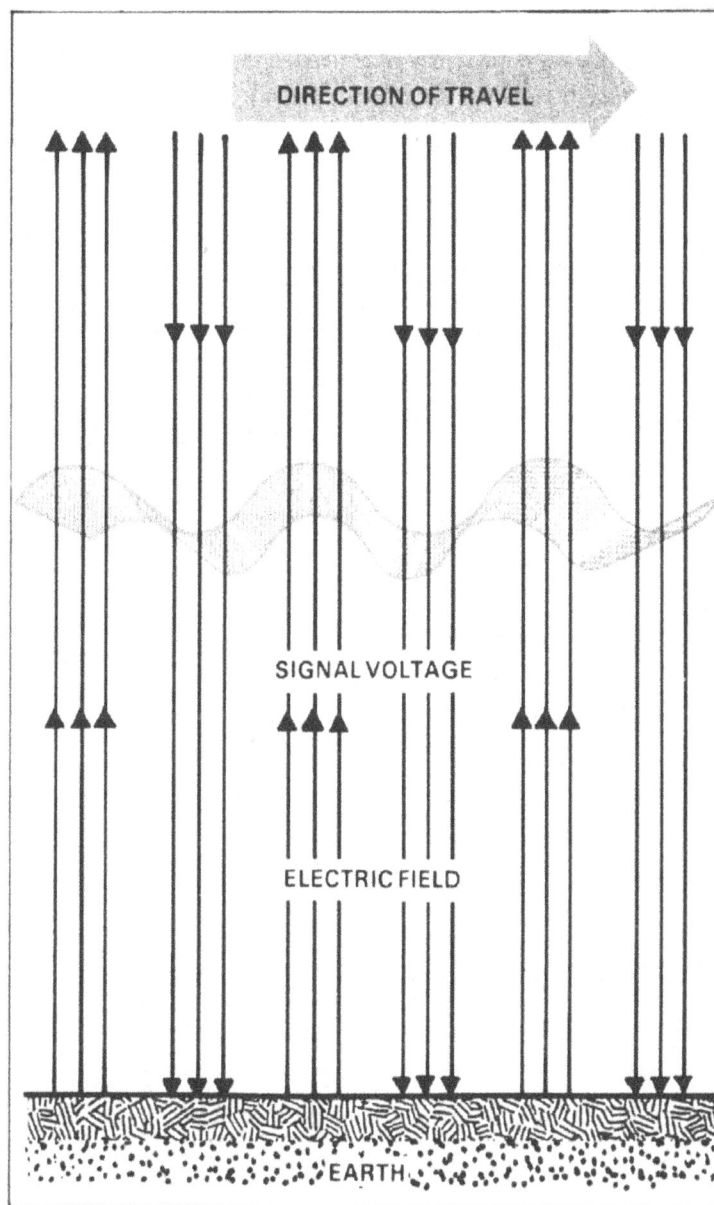

Figure 3-4. Vertically polarized signal.

Polarization Requirements for Various Frequencies.

At medium and low frequencies, ground-wave transmission is used extensively and it is necessary to use vertical polarization. Vertical lines of force are perpendicular to the ground, and the radio wave can travel a considerable distance along the ground surface with a minimum amount of loss. Because the Earth acts as a relatively good conductor at low frequencies, horizontal lines of electric force are shorted out and the useful range with the horizontal polarization is limited.

At high frequencies, with sky wave transmission, it makes little difference whether horizontal or vertical polarization is used. The sky-wave, after being reflected by the ionosphere, arrives at the receiving antenna elliptically polarized. Therefore, the transmitting and receiving antennas can be mounted either horizontally or vertically. Horizontal antennas are preferred however, since they can be made to radiate effectively at high angles and have inherent directional properties.

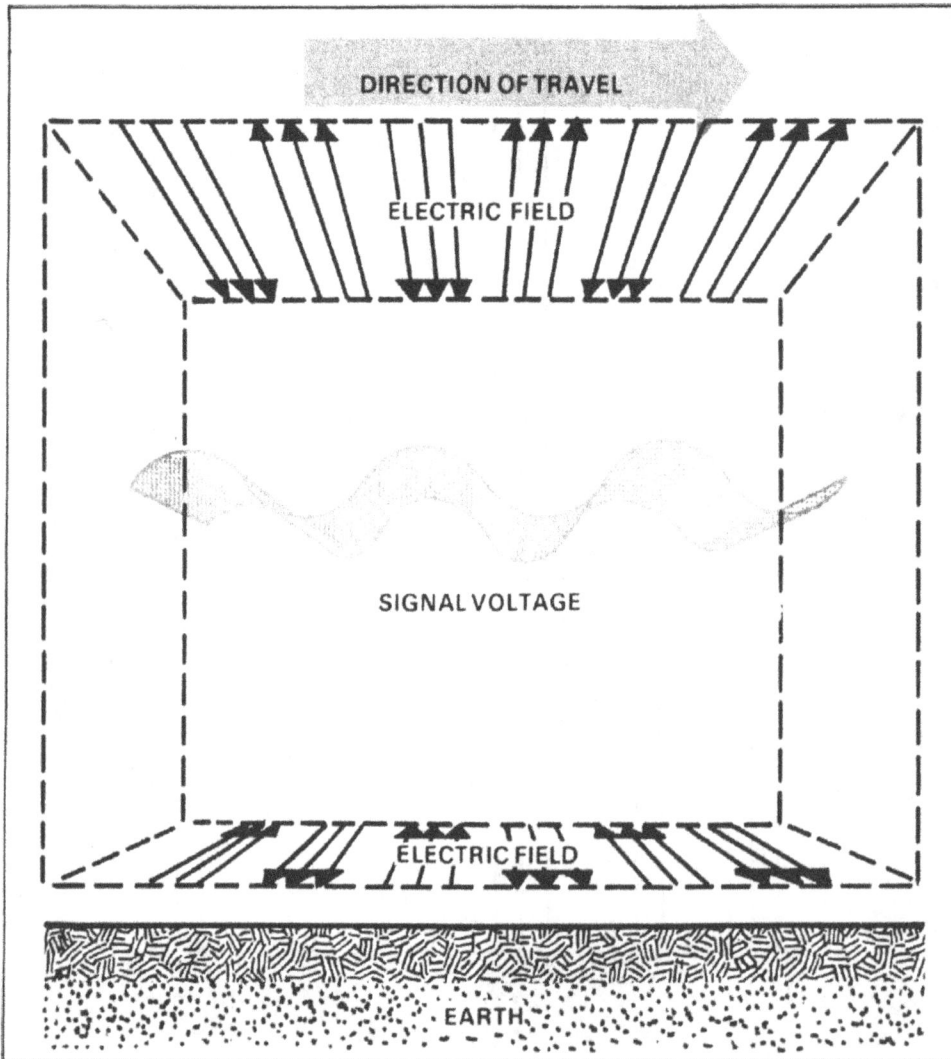

Figure 3-5. Horizontally polarized signal.

For frequencies in the very-high or ultra-high range, either horizontal or vertical polarization is satisfactory. Since the radio wave travels directly from the transmitting antenna to the receiving antenna, the original polarization produced at the transmitting antenna is maintained as the wave travels to the receiving antenna. Therefore, if a horizontal antenna is used for transmitting, a horizontal antenna must be used for receiving.

Satellites and satellite terminals use circular polarization. Circular polarization describes a wave whose plane of polarization rotates through 360° as it progresses forward. The rotation can be clockwise or counterclockwise (see fig 3-6). Circular polarization occurs when equal magnitudes of vertically and horizontally polarized waves are combined with a phase difference of 90°. Depending on their phase relationship, this causes rotation either in one direction or the other (see app L).

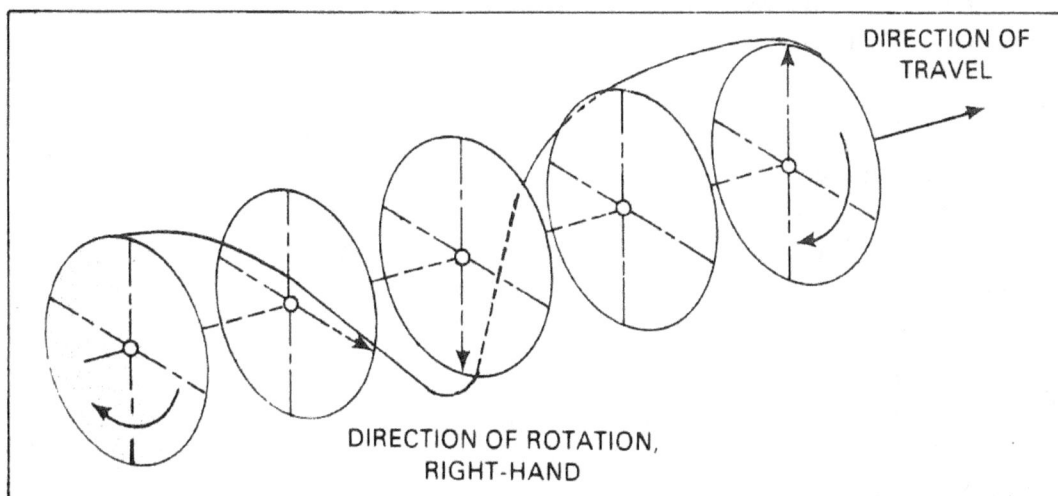

Figure 3-6. Circular polarized wave.

Advantages of Vertical Polarization.

Simple vertical half-wave and quarter-wave antennas can be used to provide omnidirectional (in all directions) communications. This is desirable in communicating with a moving vehicle. Its disadvantage is that it radiates equally to the enemy and friendly forces.

When antenna heights are limited to 3.05 meters (10 ft) or less over land, as in a vehicular installation, vertical polarization provides a stronger received signal at frequencies up to about 50 MHz. From about 50 MHz to 100 MHz, there is only a slight improvement over horizontal polarization with antennas at the same height. Above 100 MHz, the difference in signal strength between vertical and horizontal polarization is small. However, when antennas are located near dense forests, horizontally polarized waves suffer lower losses than vertically polarized waves.

Vertically polarized radiation is somewhat less affected by reflections from aircraft flying over the transmission path. With horizontal polarization, such reflections cause variations in received signal strength. An example is the picture flutter in a television set when an aircraft interferes with the transmission path. This factor is important in areas where aircraft traffic is heavy.

When vertical polarization is used, less interference is produced or picked up from strong VHF and UHF transmissions (television and FM broadcasts) because they use horizontal polarization. This factor is important when an antenna must be located in an urban area that has television or FM broadcast stations.

Advantages of Horizontal Polarization.

A simple horizontal half-wave antenna is bidirectional. This characteristic is useful in minimizing interference from certain directions.

Horizontal antennas are less likely to pick up man-made interference, which ordinarily is vertically polarized.

When antennas are located near dense forests, horizontally polarized waves suffer lower losses than vertically polarized waves, especially above 100 MHz.

Small changes in antenna location do not cause large variations in the field intensity of horizontally polarized waves when an antenna is located among trees or buildings. When vertical polarization is used, a change of only a few feet in the antenna location may have a significant effect on the received signal strength.

3-6. Directionality

Vertical receiving antennas accept radio signals equally from all horizontal directions, just as vertical transmitting antennas radiate equally in all horizontal directions. Because of this characteristic, other stations operating on the same or nearby frequencies may interfere with the desired signal and make reception difficult or impossible. However, reception of a desired signal can be improved by using directional antennas.

Horizontal half-wave antennas accept radio signals from all directions, with the strongest reception being received in a line perpendicular to the antenna (that is, broadside); and, the weakest reception being received from the direction of the ends of the antenna. Interfering signals can be eliminated or reduced by changing the antenna installation so that either end of the antenna points directly at the interfering station.

Communications over a radio circuit is satisfactory when the received signal is strong enough to override undesired signals and noise. The receiver must be within range of the transmitter. Increasing the transmitting power between two radio stations increases communications effectiveness. Also, changing the types of transmission (for example, changing from radiotelephone to CW), changing to a frequency that is not readily absorbed, or using a directional antenna aids in communications effectiveness.

Directional transmitting antennas concentrate radiation in a given direction and minimize radiation in other directions. A directional antenna may also be used to lessen interception by the enemy and interference with friendly stations.

3-7. Ground Effects

Since all practical antennas are erected over the Earth and not out in free space, except for those on satellites, the presence of the ground will alter the free space radiation patterns of antennas. The ground will also have an effect on some of the electrical characteristics of an antenna. It has the greatest effect on those antennas that must be mounted relatively close to the ground in terms of wavelength. For example, medium- and high-frequency antennas, elevated above the ground by only a fraction of a wavelength, will have radiation patterns that are quite different from the free-space patterns.

Grounded Antenna Theory.

The ground is a good conductor for medium and low frequencies and acts as a large mirror for the radiated energy. This results in the ground reflecting a large amount of energy that is radiated downward from an antenna mounted over it. Using this characteristic of the ground, an antenna only a quarter-wavelength long can be made into the equivalent of a half-wave antenna. A quarter-wave antenna erected vertically, with its lower end connected electrically to the ground (fig 3-7), behaves like a half-wave antenna. Under these conditions, the ground takes the place of the missing quarter-wavelength, and the reflections supply that part of the radiated energy that normally would be supplied by the lower half of an ungrounded half-wave antenna.

Types of Grounds.

When grounded antennas are used, it is especially important that the ground has as high a conductivity as possible. This reduces ground losses and provides the best possible reflecting surface for the down-going radiated energy from the antenna. At low and medium frequencies, the ground acts as a sufficiently good conductor. Therefore, the ground connection must be made in such a way as to introduce the least possible amount of resistance to ground. At higher frequencies, artificial grounds constructed of large metal surfaces are common.

The ground connections take many forms, depending on the type of installation and the loss that can be tolerated. In many simple field installations, the ground connection is made by means of one or more metal rods driven into the soil. Where more satisfactory arrangements cannot be made, ground leads can be connected to existing devices which are grounded. Metal structures or underground pipe systems are commonly used as ground connections. In an emergency, a ground connection can be made by forcing one or more bayonets into the soil.

When an antenna must be erected over soil with low conductivity, treat the soil to reduce its resistance. The soil should be treated with substances that are highly conductive when in solution. Some of these substances, listed in order of preference, are sodium chloride (common salt), calcium chloride, copper sulphate (blue vitriol), magnesium sulphate (Epsom salt, and potassium nitrate (saltpeter). The amount required depends on the type of soil and its moisture content.

> *WARNING: When these substances are used, it is important that they do not get into nearby drinking water supplies.*

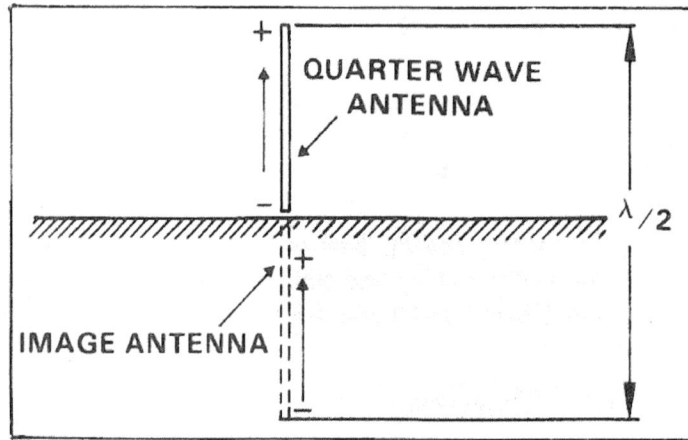

Figure 3-7. Quarter-wave antenna connected to ground.

For simple installations, a single ground rod can be fabricated in the field from pipe or conduit. It is important that a low resistance connection be made between the ground wire and the ground rod. The rod should be cleaned thoroughly by scraping and sandpapering at the point where the connection is to be made, and a clean ground clamp should be installed. A ground wire can then be soldered or joined to the clamp. This joint should be covered with tape to prevent an increase in resistance because of oxidation.

Counterpoise.

When an actual ground connection cannot be used because of the high resistance of the soil or because a large buried ground system is not practical, a counterpoise may be used to replace the usual direct ground connection. The counterpoise (fig 3-8) consists of a device made of wire which is erected a short distance above the ground and insulated from it. The size of the counterpoise should be at least equal to or larger than the size of the antenna.

When the antenna is mounted vertically, the counterpoise should be made into a simple geometric pattern. Perfect symmetry is not required. The counterpoise appears to the antenna as an artificial ground that helps to produce the required radiation pattern.

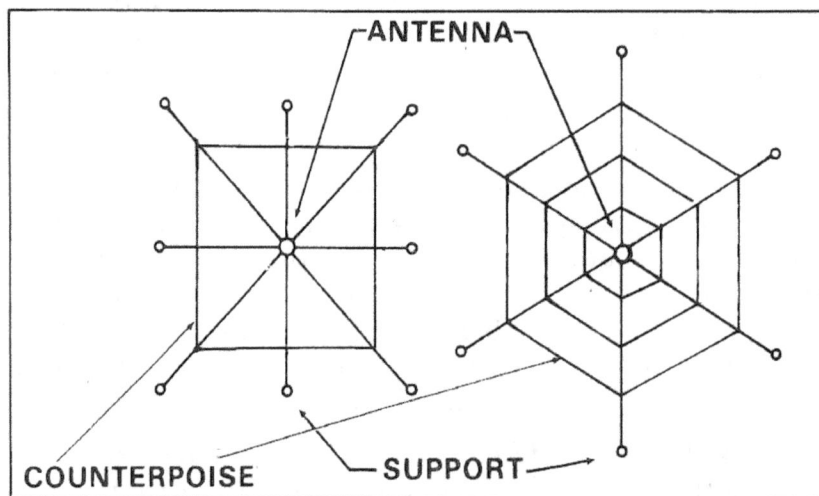

Figure 3-8. Wire counterpoise.

In some VHF antenna installations on vehicles, the metal roof of the vehicle (or shelter) is used as a counterpoise for the antenna.

Small counterpoises of metal mesh are sometimes used with special VHF antennas that must be located a considerable distance above the ground.

Ground Screen.

A ground screen consists of a fairly large area of metal mesh or screen that is laid on the surface of the ground under the antenna. There are two specific advantages in using ground screens. First, the ground screen reduces ground absorption losses that occur when an antenna is erected over ground with poor conductivity. Second, the height of the antenna can be set accurately. As a result of this, the radiation resistance of the antenna can be determined more accurately (See TM 11-666, para 61).

3-8. Antenna Length

The length of an antenna must be considered in two ways. It has both a physical and an electrical length, and the two are never the same. The reduced velocity of the wave on the antenna and a capacitive effect (known as *end effect*) make the antenna seem longer electrically than it is physically. The contributing factors are the ratio of the diameter of the antenna to its length and the capacitive effect of terminal equipment (insulators, clamps, etc.) used to support the antenna.

To calculate the physical length of an antenna, use a correction of 0.95 for frequencies between 3.0 and 50.0 MHz. The figures given below are for a half-wave antenna.

$$\text{Length (meters)} = \frac{150 \times 0.95}{\text{Frequency in MHz}} = \frac{142.5}{\text{Frequency in MHz}}$$

$$\text{Length (feet)} = \frac{492 \times 0.95}{\text{Frequency in MHz}} = \frac{468}{\text{Frequency in MHz}}$$

The length of a long-wire antenna (one wavelength or longer) for harmonic operation is calculated by using the following formula.

$$\text{Length (meters)} = \frac{150 (N-0.05)}{\text{Freq MHz}}$$

$$\text{Length (feet)} = \frac{492 (N-0.05)}{\text{Freq MHz}}$$

Where N = number of half-wavelengths in the total length of the antenna.

For example, if the number of half-wavelengths is 3 and the frequency in MHz is 7, then:

$$\text{Length (meters)} = \frac{150(N-0.05)}{\text{Freq MHz}} = \frac{150(3-.05)}{7} = \frac{150 \times 2.95}{7} = \frac{442.50}{7} = 63.2 \text{ meters}$$

3-9. Antenna Orientation
Azimuth.

If the azimuth of the radio path is not provided, the azimuth should be determined by the best available means. The accuracy required in determining the azimuth of the path is dependent upon the radiation pattern of the directional antenna. If the antenna beam width is very wide (for example, 90° angle between half-power points, fig 3-9), an error of 10° in azimuth is of little consequence. In transportable operation, the rhombic and V antennas may have such a narrow beam as to require great accuracy in azimuth determination. The antenna should be erected for the correct azimuth. Great accuracy is not required in erecting broad-beam antennas. Unless a line of known azimuth is available at the site, the direction of the path is best determined by a magnetic compass. Figure 3-10 is a map of magnetic declination, showing the variation of the compass needle from the true north. When the compass is held so that the needle points to the direction indicated for the location on the map, all directions indicated by the compass will be true.

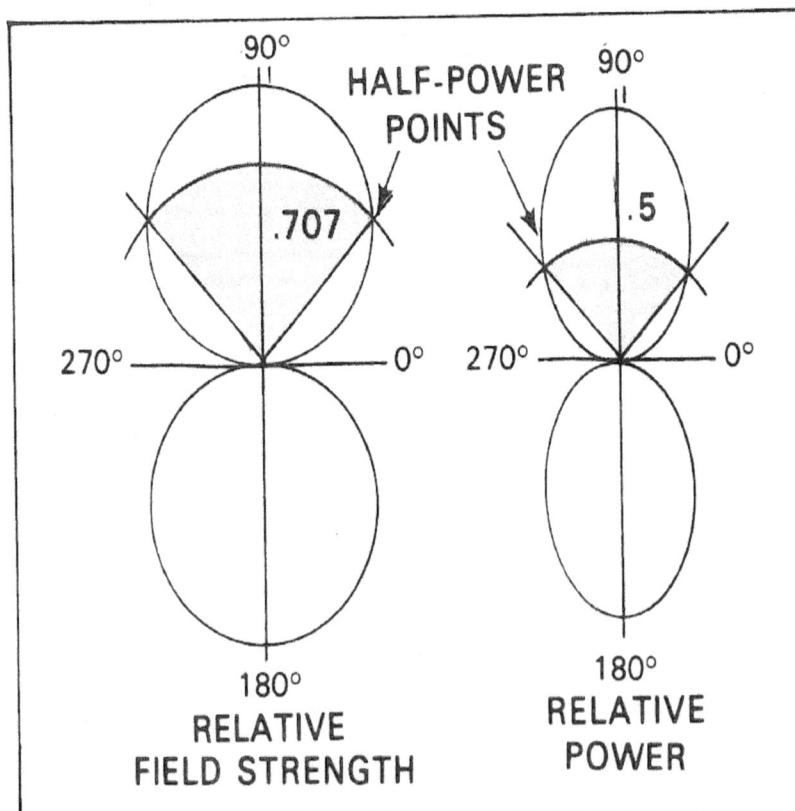

Figure 3-9. Beam width measured on relative field strength and relative power patterns.

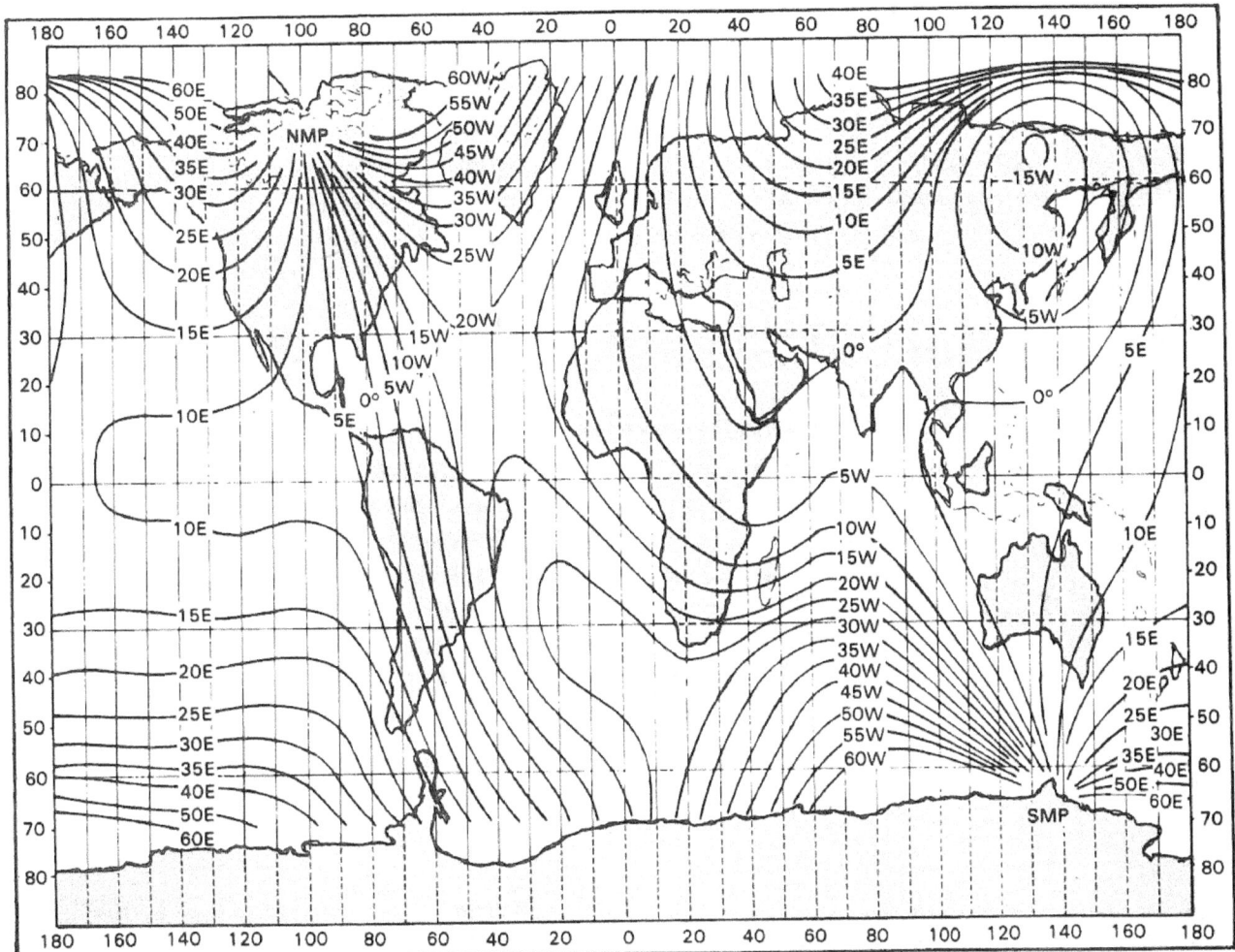

Figure 3-10. Magnetic declination over the world.

Improvement of Marginal Communications.

Under certain situations, it may not be feasible to orient directional antennas to the correct azimuth of the desired radio path. As a result, marginal communications may suffer. To improve marginal communications, follow the procedure presented below.

- Check, tighten, and tape cable couplings and connections.

- Retune all transmitters and receivers in the circuit.

- Check to see that antennas are adjusted for the proper operating frequency.

- Try changing the heights of antennas.

- Try moving the antenna a short distance away and in different locations from its original location.

- Separate transmitters from receiving equipment, if feasible.

Transmission and Reception of Strong Signals.

After an adequate site has been selected and the proper antenna orientation obtained, the signal level at the receiver will be proportional to the strength of the transmitted signal.

If a high-gain antenna is used, a stronger signal can be obtained. Losses between the antenna and the equipment can be reduced by using a high quality transmission line, as short as possible, and properly matched at both ends.

WARNING: Excessive signal strength may result in enemy intercept and interference or in your interfering with adjacent frequencies.

Section III. Types of Antennas

3-10. Tactical Considerations

Tactical antennas are specially designed to be rugged and permit mobility with the least possible sacrifice of efficiency. They are also designed to take abuse. Some are mounted on the sides of vehicles that have to move over rough terrain; others are mounted on tops of single masts or suspended between sets of masts. The smallest antennas are mounted on the helmets of personnel who use the radio sets. All tactical antennas must be easy to install. Large ones must be easy to take apart and pack and they must be easy to transport.

Several types of transmitting and receiving antennas are shown in figure 3-11.

- A of the figure is a rhombic antenna.
- B is a half-wave Hertz antenna.
- C is an end-fed, vertical antenna, also called a whip antenna.
- D is a loop antenna that receives a strong signal in directions as shown and almost no signal in other directions.
- E is an antenna group OE-254/GRC which is an omnidirectional, biconical antenna designed for broadband operation.
- F is a long-wire antenna.
- G is a vertical half-rhombic antenna.
- H is a directional half-rhombic antenna.

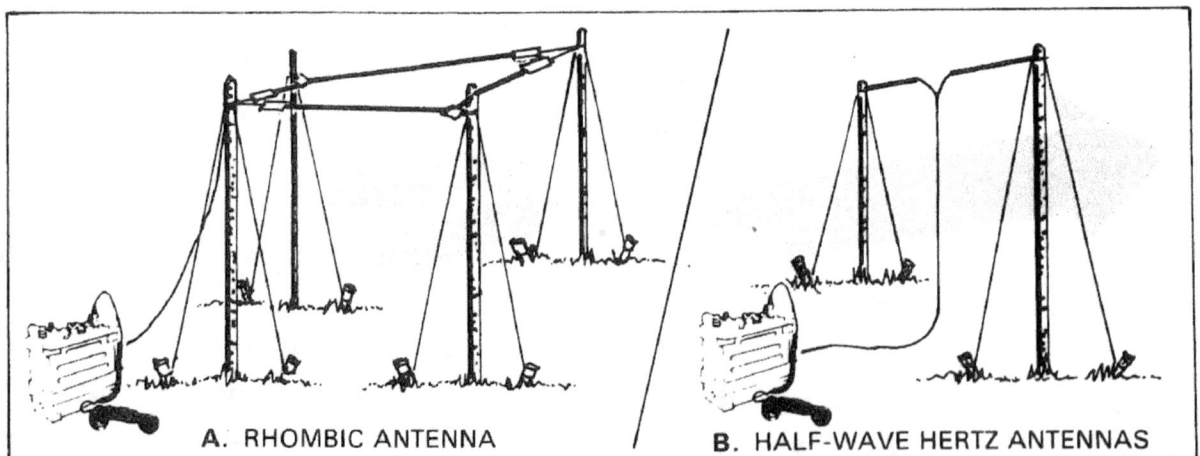

A. RHOMBIC ANTENNA   B. HALF-WAVE HERTZ ANTENNAS

Figure 3-11. Types of antennas.

NEAR-VERTICAL INCIDENCE SKY-WAVE PROPAGATION CONCEPT

M-1. Evaluation of Communications Techniques

The standard communications techniques used in the past will not support the widely deployed and the fast-moving formations we intend to use to counter the modern threat. Coupling this with the problems that can be expected in deploying multichannel LOS systems with relays to keep up with present and future operation, high frequency (HF) radio and the near-vertical incidence sky-wave (NVIS) mode take on new importance. High frequency radio is quickly deployable, securable, and capable of data transmission. It will be the first, and frequently the only, means of communicating with fast-moving or widely separated units. It may also provide the first long-range system to recover from a nuclear attack. With this reliance on HF radio, communications planners, commanders, and operators must be familiar with NVIS techniques and their applications and shortcomings in order to provide more reliable communications.

M-2. Problems Encountered in Propagation of Radio Waves

Under ideal conditions, ground wave component of a radio wave becomes unusable at about 80 kilometers (50 mi) (fig 2-12). Under actual field conditions, this range can be much less, sometimes as little as 3 kilometers (2 mi). Sky waves, generated by standard antennas (for example, doublets) which efficiently launch the sky wave, will not return to earth at a range of less than 161 kilometers (100 mi). This can leave a skip zone of at least 80 to 113 kilometers (50 to 70 mi) where HF communications will not function. This means that units such as long-range patrols, armored cavalry deployed as advance or covering forces, air defense early warning teams, and many division-corps, division-brigade, division-DISCOM and division-DIVARTY stations are in the skip zone and thus unreachable by HF radio even though HF is a primary means of communication to these units.

M-3. Concept of Near-Vertical Incidence Sky-Wave Radiation

Energy radiated in a near-vertical incidence direction is not reflected down to a pinpoint on the Earth's surface. If it is radiated on too high a frequency, the energy penetrates the ionosphere and continues on out into space. Energy radiated on a low enough frequency is reflected back to earth at all angles (including the zenith), resulting in the energy striking the earth in an omnidirectional pattern without dead spots (that is, without a skip zone). Such a mode is called a near-vertical incidence sky wave (NVIS). The concept is illustrated in figure M-1.

This effect is similar to taking a hose with a fog nozzle and pointing it straight up. The water falling back to earth covers a circular pattern continuously out to a given distance. A typical receive signal pattern for antenna AS-2259/GR is shown in figure M-2, and the path length and incident angle are shown in figure M-3. A typical elevation plane pattern is shown in figure M-4. The main difference between this short-range NVIS mode and the standard long-range sky-wave HF mode is the lower frequency required to avoid penetrating the ionosphere and the angle of incident signal upon the ionosphere. In order to attain a NVIS effect, the energy must be radiated strong enough at angles greater than about 75 or 80 degrees from the horizontal on a frequency that the ionosphere will reflect at that location and time. The ionospheric layers will reflect this energy in an umbrella-type pattern with no skip zone. Any ground wave present with the NVIS signal will result in undesirable wave interference effects (such as, fading) if the amplitudes are comparable. However, proper antenna selection will reduce ground-wave radiated energy to a minimum, and this will reduce the fading problems. Ranges for the NVIS mode are shown in figure M-3 for typical ionosphere height and take-off angles. Since NVIS paths are purely sky wave, the path losses are nearly constant at about 110 dB ±10 dB. Relative gain performance of the AS-2259/GR NVIS antenna is shown in figure M-5. This is significant for the tactical communicator because all the energy arriving at the receiving antenna is coming from above at about the same strength over all of the communications ranges of interest. This means the effect of terrain and vegetation (when operating from defiladed positions such as valleys) are greatly reduced, and the receive signal strength will not vary greatly.

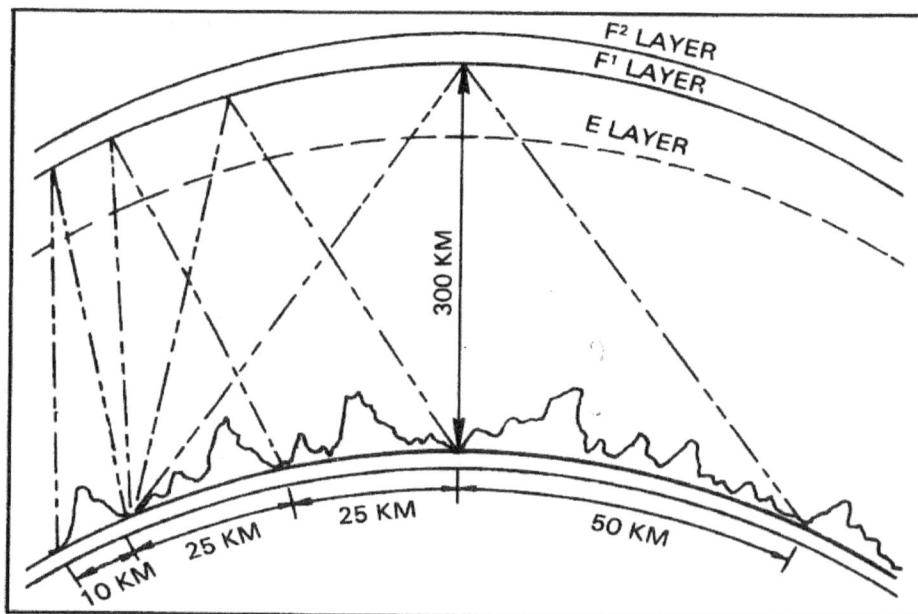

Figure M-1. Near-vertical incidence sky-wave propagation concept.

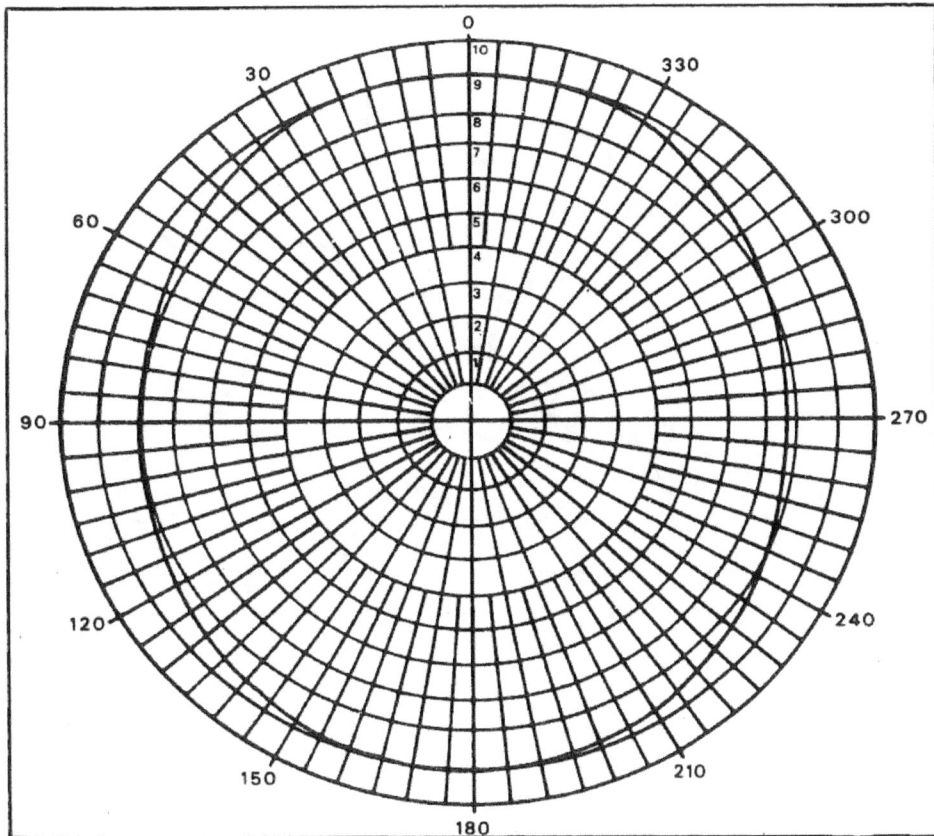

Figure M-2. Near-vertical incidence sky-wave antenna
typical azimuth plane pattern.

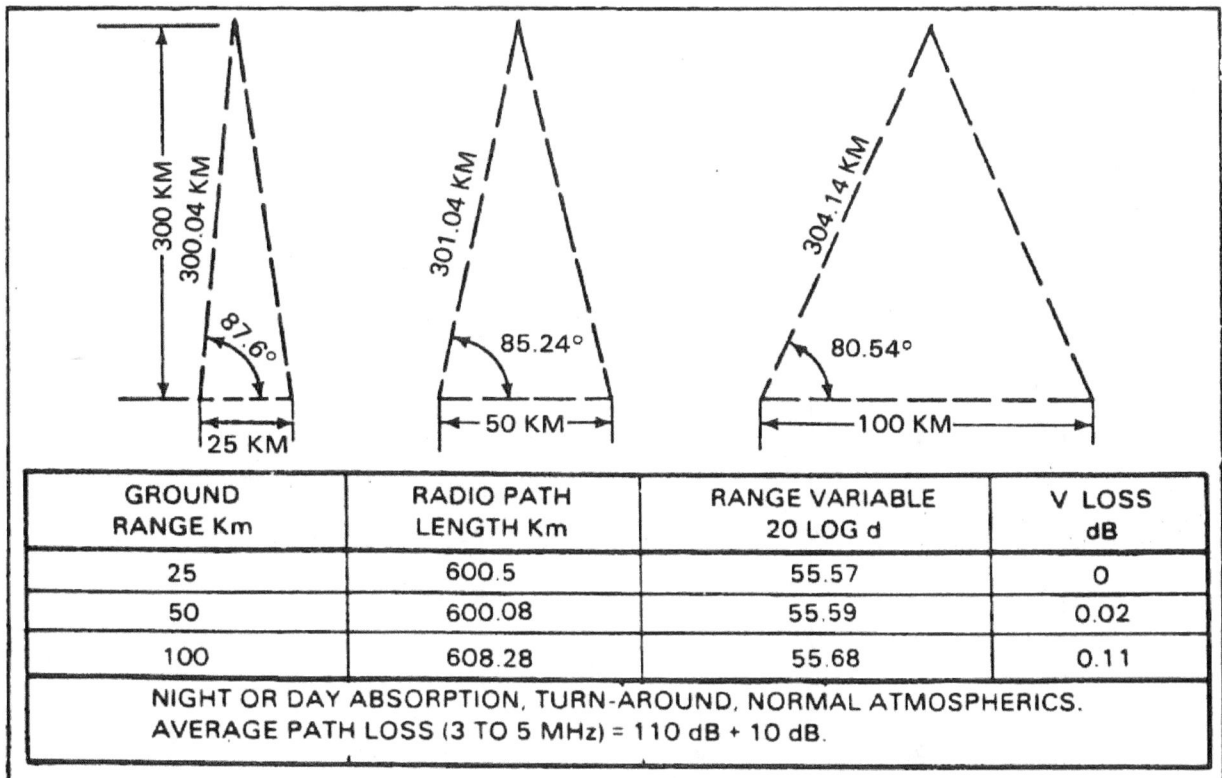

GROUND RANGE Km	RADIO PATH LENGTH Km	RANGE VARIABLE 20 LOG d	V LOSS dB
25	600.5	55.57	0
50	600.08	55.59	0.02
100	608.28	55.68	0.11

NIGHT OR DAY ABSORPTION, TURN-AROUND, NORMAL ATMOSPHERICS.
AVERAGE PATH LOSS (3 TO 5 MHz) = 110 dB + 10 dB.

Figure M-3. Path length and incident angle (near-vertical incidence sky-wave
mode).

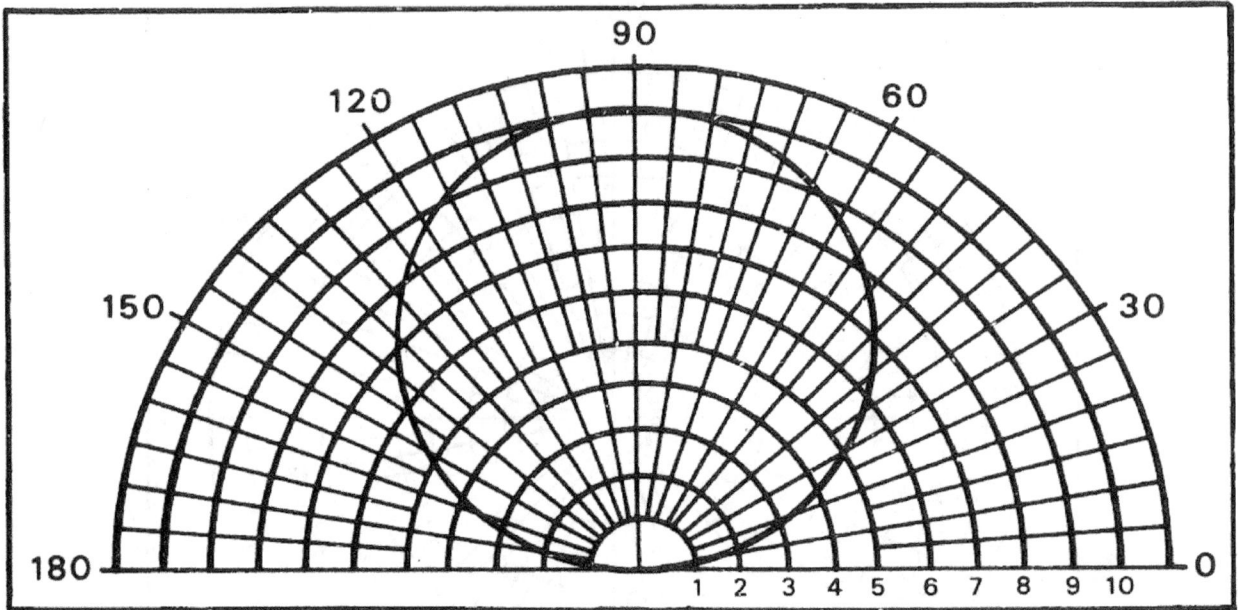

Figure M-4. Typical elevation plane pattern.

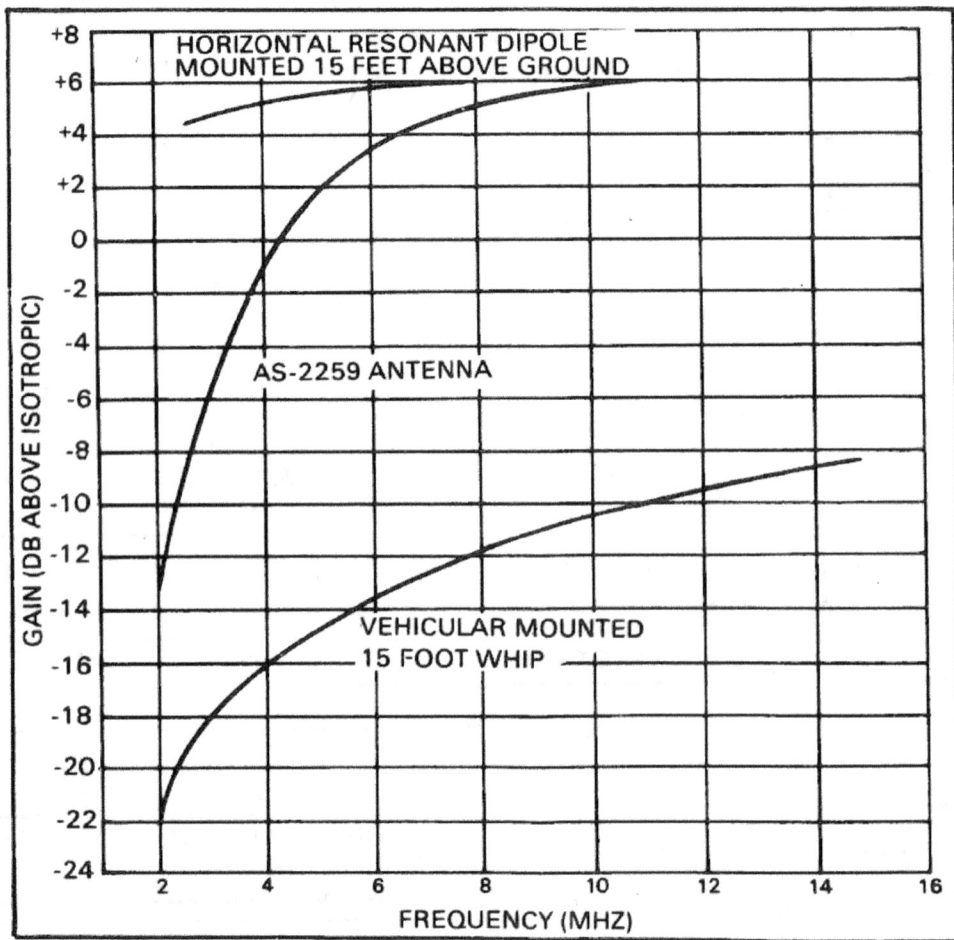

Figure M-5. Relative gain performance of AS-2259 antenna.

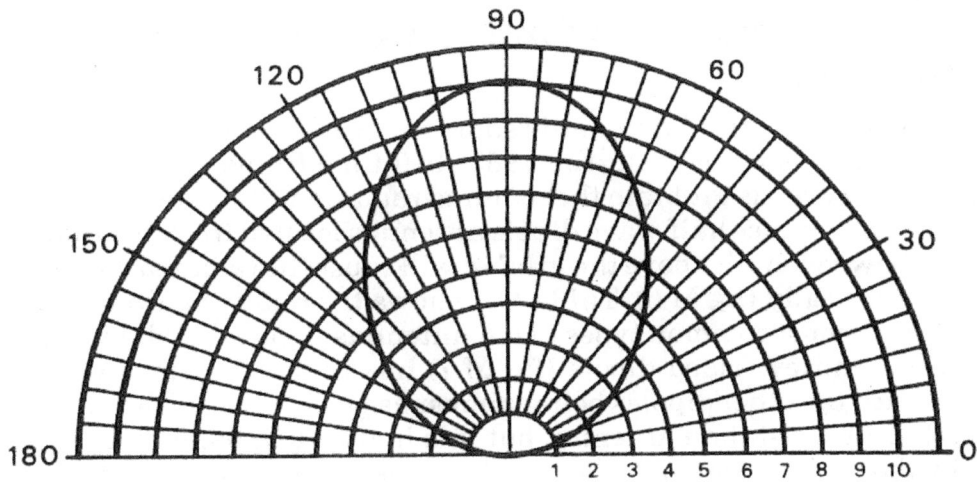

Figure M-6 Typical elevation plane patterns for half-wavelength antennas one-eighth wavelength or less above ground.

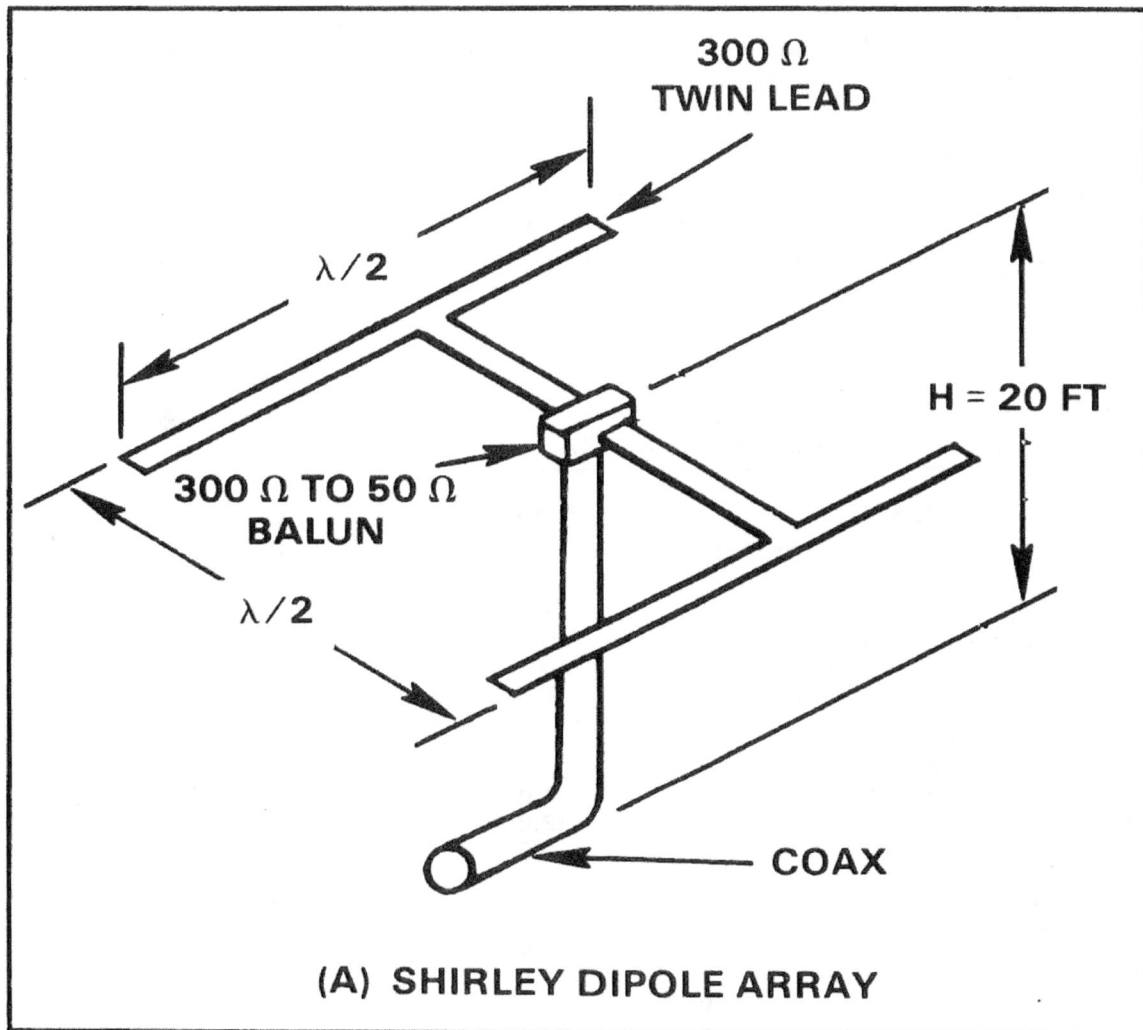

(A) SHIRLEY DIPOLE ARRAY

Figure M-7. Half-wave Shirley folded dipole.

M-4. Assessment of Characteristics of Common Antennas

It is obvious that the Army needs short-range HF communications in the 2-30 MHz frequency band in the 1985-1990 time frame and beyond. The problem, however, is to obtain the required radiation characteristics. This is not difficult, because half-wave dipole antennas located from one-quarter to one-tenth wavelength above the ground will cause the radiated energy to be directed vertically (fig M-6). Table M-1 shows the relative gain toward the zenith of the most common types of HF antennas. This table shows that the half-wave Shirley folded dipole (fig M-7) has the most gain towards the zenith (with the other dipoles being almost as good). The Shirley dipole is a good NVIS base station antenna, but it is limited to a band of frequencies within about 10 percent of the design frequency. The fan dipole (fig M-8 and table M-1) performs almost as well, and it provides more frequency flexibility (for example, day, night, and transition period frequencies). For tactical communications, these dipoles can be easily deployed in a field expedient manner because they can be located close to the ground. For mobile or shoot-and-scoot type operations, vehicular-mounted antennas are required. This is the standard 5-meter (16 1/2-foot) whip bent down to a horizontal position (fig M-9). In this configuration, the whip is essentially an asymmetrical dipole (with the vehicle body forming one side) located close to the Earth. A significant amount of energy is directed upward (fig M-6 for typical pattern) to be reflected back by the ionosphere in an umbrella pattern. For use, while operating on the move, the whip antenna must be tied across or parallel to the vehicle or shelter. This configuration is like an asymmetrical open-wire line, and it also directs some energy upwards although with less efficiency. There are still no skip zones, but received signal levels are weaker than with the whip tied back as shown in figure M-9.

Table M-1. Summary of relative gain toward the zenith for field-expedient high frequency antennas (in dB)

ANTENNA	CLEARING	75-FT FOREST	50-FT FOREST
$\lambda/2$ Unbalanced Single-Wire Dipole	+1.0	-2.8	-1.2, -1.7
$\lambda/2$ Balanced Single-Wire Dipole	+0.5	-3.7	No data
$\lambda/2$ Folded Dipole (300:50 ft balun)	+0.2	-1.0	No data
$\lambda/4$ Short (Loaded to $\lambda/2$) Dipole	-3.0	-5.2	No data
$\lambda/2$ Sleeve Dipole (on ground)	-32.1	-28.3	No data
3-Freq. Fan Dipole @ 15 ft	-0.4	-5.1	No data
3-Freq. Fan Dipole @ 12 ft	-2.4	-5.0	No data
3-Freq. Fan Dipole @ 9 ft	-4.0	-8.1	No data
Shirley Folded Dipole	+3.0	-0.3	No data
$3\lambda/4$ Inverted L (1:h — 2:1)	-0.0	-2.8	No data
$3\lambda/4$ Inverted L (1:h — 3:1)	-0.8	-3.3	No data
$3\lambda/4$ Inverted L (1:h — 4:1)	-1.0	-5.8	No data
$3\lambda/4$ Inverted L (1:h — 5:1)	-2.0	-0.3	-10.7, -12.5
30° Slant Wire ($\lambda/4$ elevated)	-10.1	-14.8	-13.5, -14.2
60° Slant Wire ($\lambda/4$ elevated)	-11.8	-14.8	No Data
10-ft Square (Vertical Plane) Loop @ 6 ft	-21.1	-25.3	No data
16.5-ft Whip	-41.5	-44.0	-25.0, -25.2

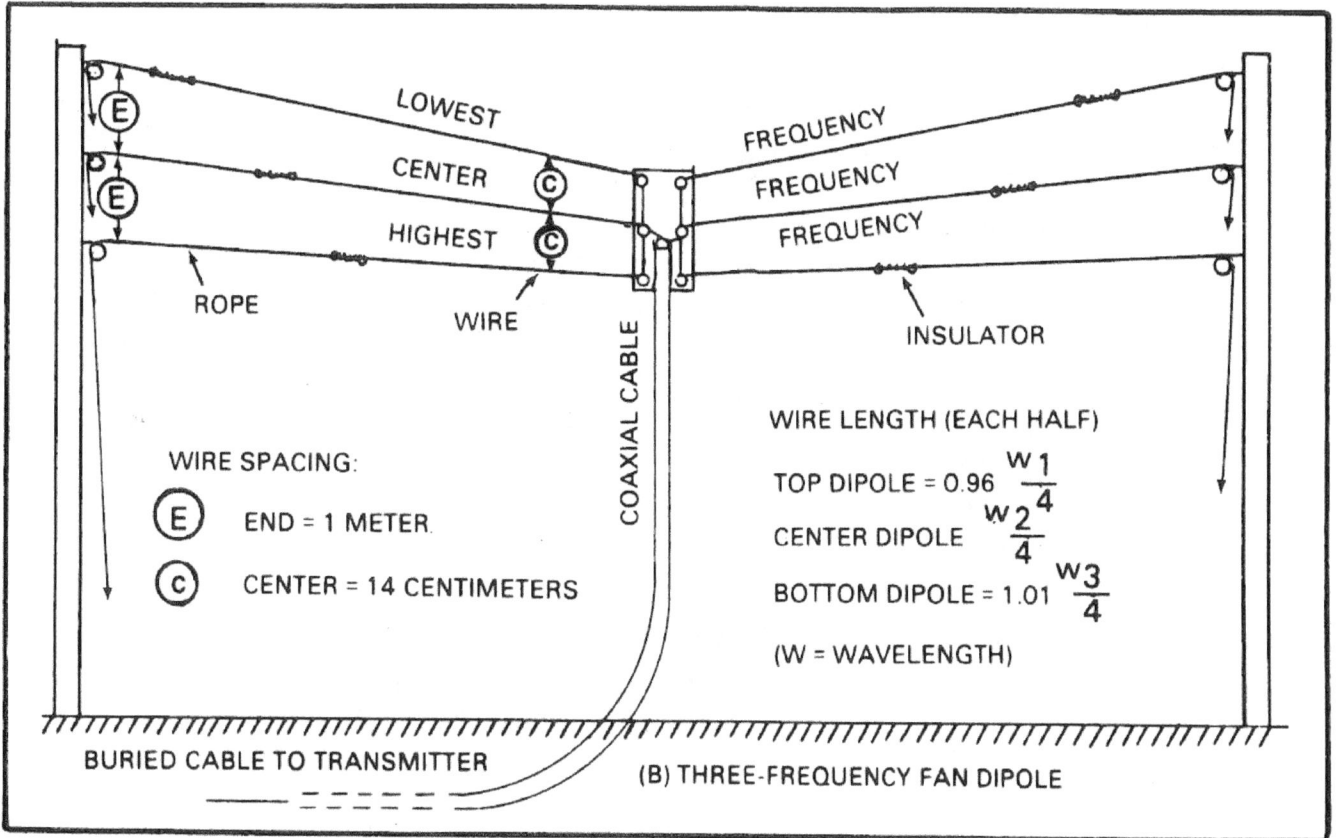

LOWEST FREQUENCY

CENTER FREQUENCY

HIGHEST FREQUENCY

ROPE

WIRE

COAXIAL CABLE

INSULATOR

WIRE SPACING:

(E) END = 1 METER

(C) CENTER = 14 CENTIMETERS

WIRE LENGTH (EACH HALF)

TOP DIPOLE = $0.96 \frac{W_1}{4}$

CENTER DIPOLE $\frac{W_2}{4}$

BOTTOM DIPOLE = $1.01 \frac{W_3}{4}$

(W = WAVELENGTH)

BURIED CABLE TO TRANSMITTER

(B) THREE-FREQUENCY FAN DIPOLE

Figure M-8. Fan dipole NVIS base station antenna.

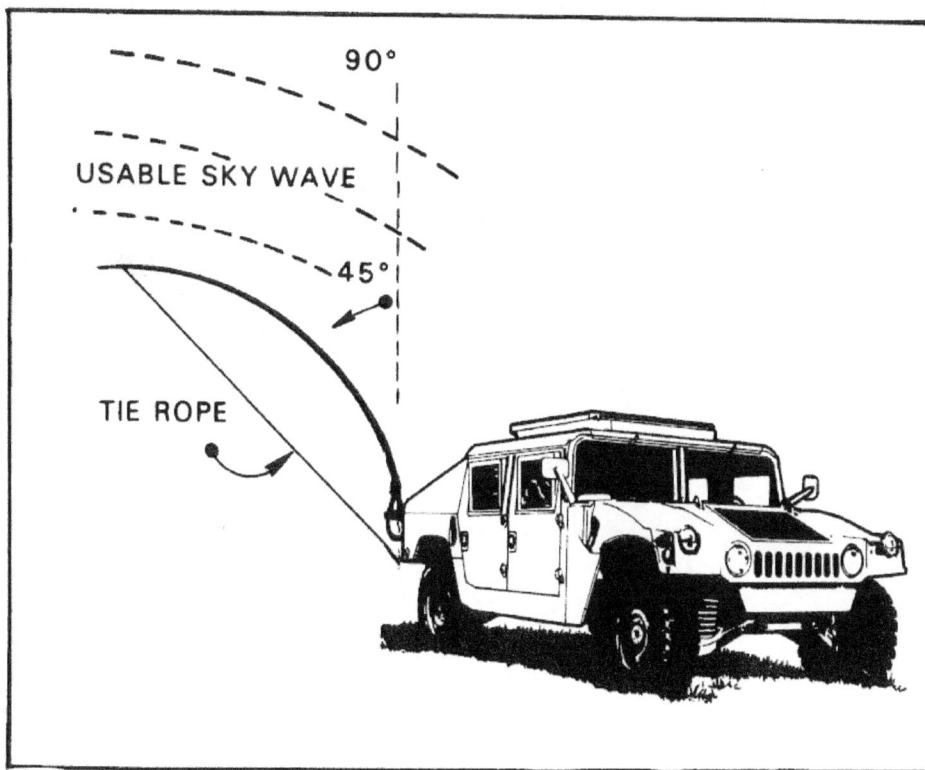

90°

USABLE SKY WAVE

45°

TIE ROPE

Figure M-9. Tying the whip antenna down.

M-5. Orientation of Antenna

Wire dipole antennas have always been sited so that the broadside of the antenna was pointed toward the receiving station(s). This is still the correct approach for long-haul paths. This antenna orientation is not necessary when using the NVIS mode. For NVIS operation, antenna orientation does not matter since all the energy is directed upward and returns to earth in an omnidirectional pattern. This means that the dipole should be erected at any orientation that is convenient at the particular radio site without regard to the location of other stations. This holds true except when operating in the region of the magnetic dip equator (fig M-10). When operating near the dip equator (such as, within 500 kilometers (311 mi)), the dipole antennas should be oriented in a magnetically north-south direction for greater receive signal levels for all NVIS path bearings. Antenna orientation broadside to the path direction must be retained near the dip equator and elsewhere for longer sky-wave paths.

M-6. Problems in Using the NVIS Concept

While use of the NVIS technique does provide beyond line-of-sight, skip-zone-free communications, there are some drawbacks in its use that must be understood in order to minimize them.

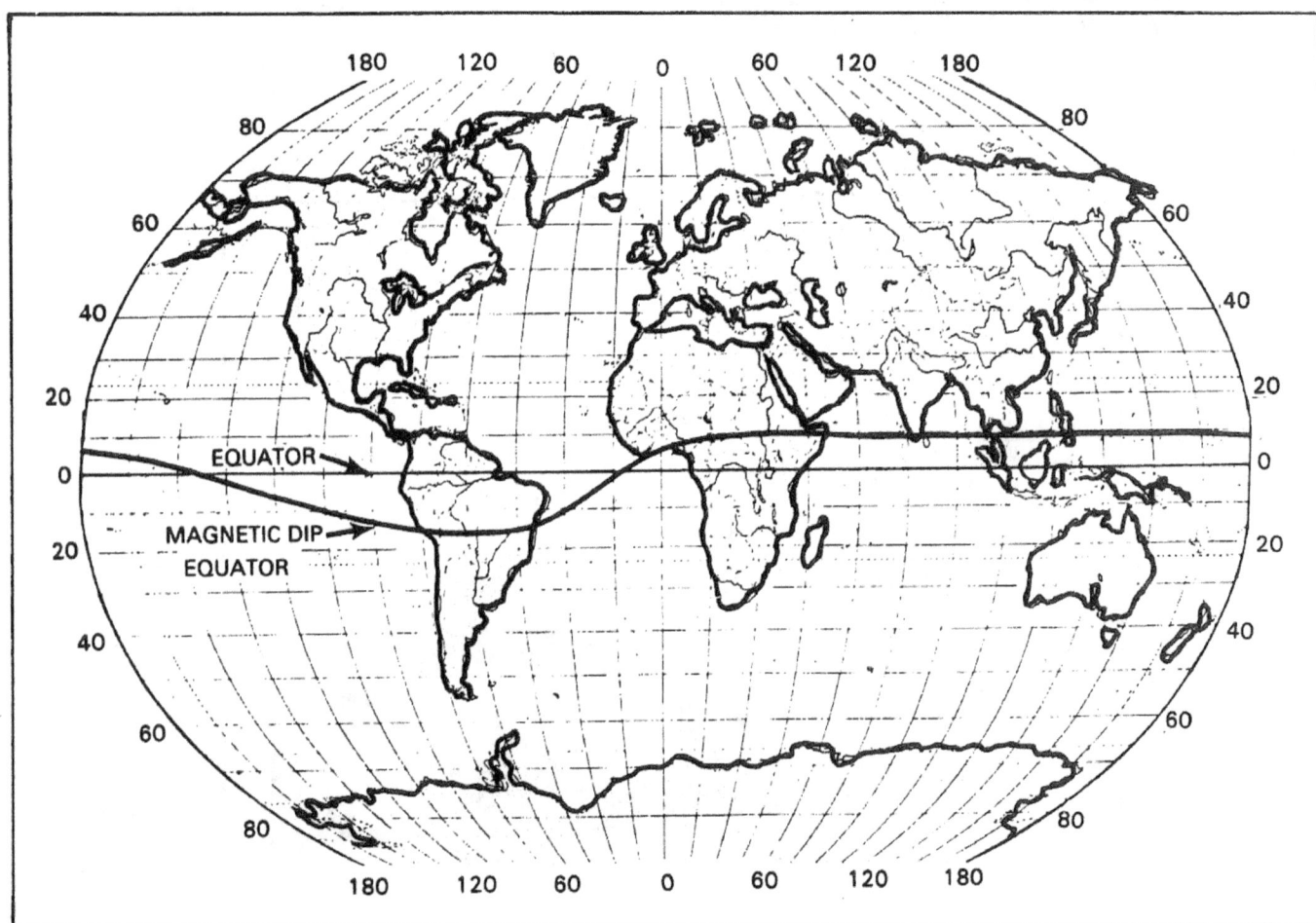

Figure M-10. Magnetic dip equator.

Interference Between Ground Wave and Sky Wave.

Where both a NVIS and ground-wave signal are present, the ground wave can cause destructive interference. Proper antenna selection will suppress ground-wave radiation and minimize this effect while maximizing the amount of energy going into the NVIS mode.

High Take-Off Angles.

In order to produce radiation which is nearly vertical, antennas must be selected and located carefully in order to minimize the ground-wave radiation and maximize the energy radiated towards the zenith. This can be accomplished by using specially designed antennas such as AS-2259/GR or by locating standard dipole (doublet) antennas one-quarter to one-tenth wavelength from the ground in order to direct the energy toward the zenith (fig M-11). A typical measured dipole pattern (power gain) is shown in figure M-12.

Critical Frequency Selection.

As in all sky-wave propagation, there is a critical frequency (fo) above which radiated energy will not be reflected by the ionosphere but will pass through it (TM 11-666). This frequency is related approximately to the angle of incidence.

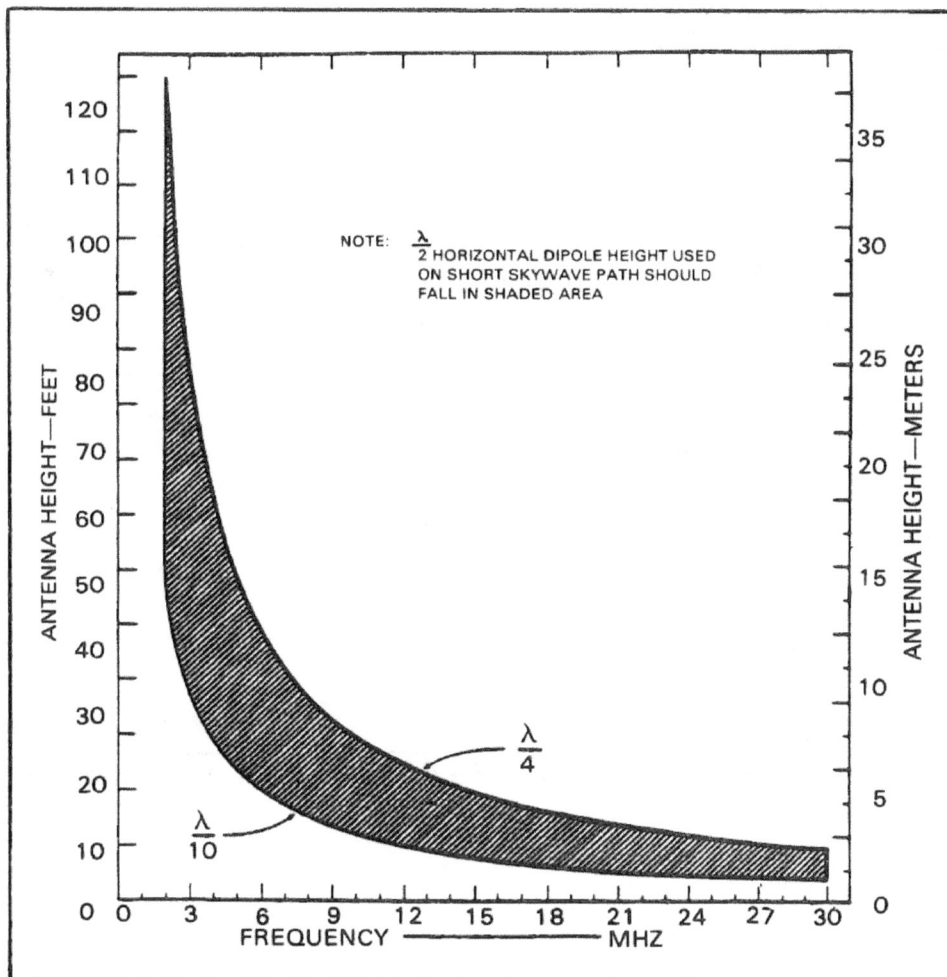

Figure M-11. Recommended dipole height for NVIS applications.

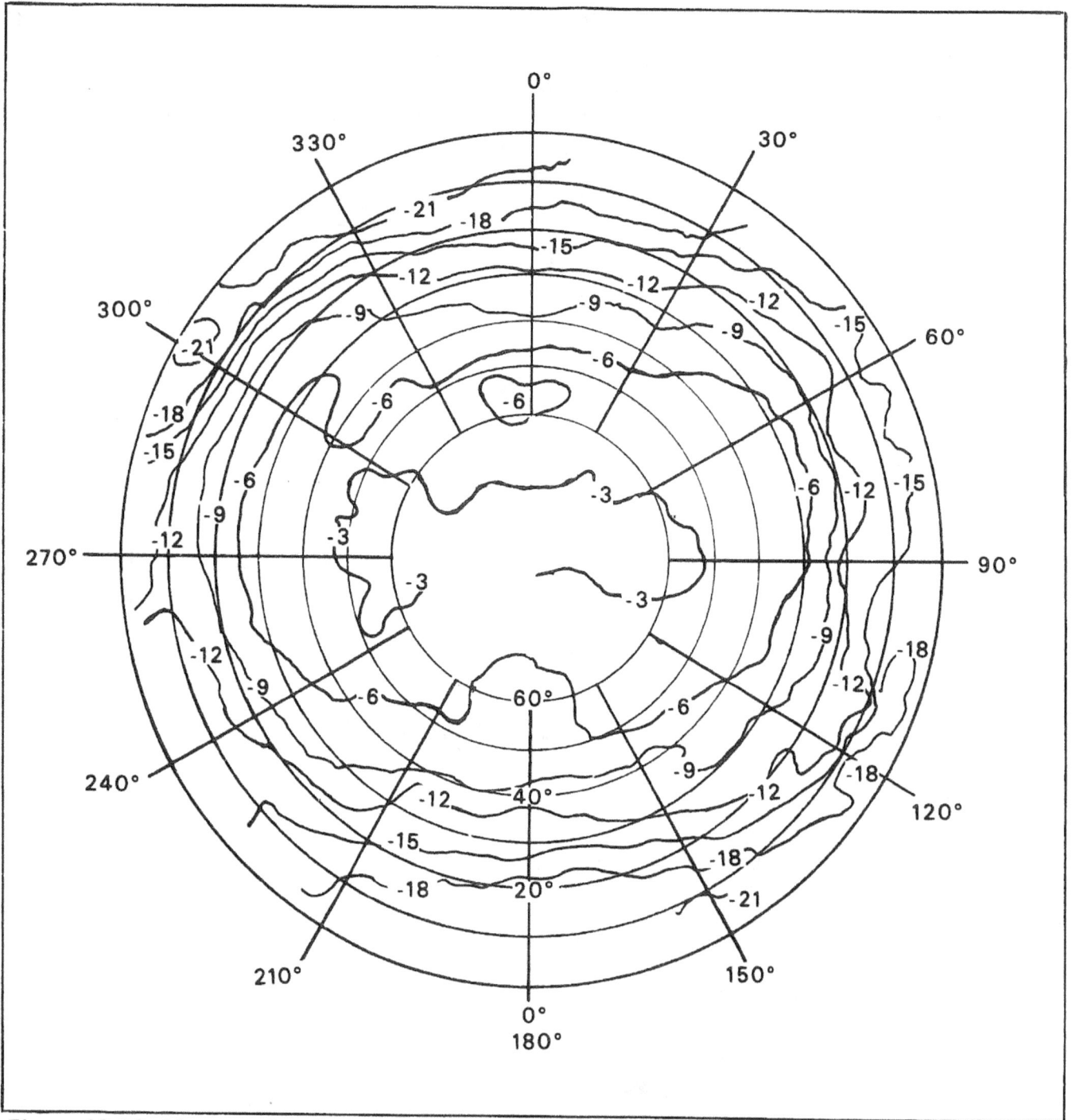

Figure M-12. Measured radiation pattern of the 8-MHz 23-foot high unbalanced dipole.

This means that the useful frequency range varies in accordance with the path length. The shorter the path, the lower the MUF and the smaller the frequency range. In practice, this limits the NVIS mode of operation to the 2-to 4-MHz range at night and to the 4- to 8-MHz range during the day (fig M-13). These nominal limits will vary with the 11-year sunspot cycle and they will be smaller during sunspot minimums (for example, 1985-86). This restriction of the frequency range is due to the physics of the situation and cannot be overcome. Some problems can be expected when operating on the NVIS mode in this portion of the HF spectrum.

The range of frequencies between the MUF and the LUF is limited, and frequency assignment may be a problem.

The lower portion of the band which supports NVIS is somewhat congested with aviation, marine, broadcast, and amateur radio which limits frequencies available.

Atmospheric noise is higher in this portion of the HF spectrum in the afternoon and night.

Man-made noise tends to be higher in this portion of the HF spectrum.

M-7. Advantages in Using the NVIS Concept

After the foregoing problems are overcome, there are many advantages in using the NVIS concept.

The tactical environment.

● There are skip-zone-free omnidirectional communications.

● Terrain does not effect loss of signal. This gives a more constant received signal level over the operational range instead of one which varies widely with distance.

● Operators are able to operate from protected, dug-in positions. Thus tactical commanders do not have to control the high ground for HF communications purposes.

● Orientation of doublets and inverted antennas become noncritical.

The EW environment

● **There is a lower probability of geolocation.** NVIS energy is received from above at very steep angles, which makes direction finding (DF) from nearby (but beyond ground-wave range) locations more difficult.

● **Communications are harder to jam.** Ground-wave jammers are subject to path loss. Terrain features can be used to attenuate a ground wave jammer without degrading the desired communication path. The jamming signal will be attenuated by terrain, while the sky-wave NVIS path loss will be constant. This will force the jammer to move very close to the target or put out more power. Either tactic makes jamming more difficult.

● **Operators can use low-power successfully.** The NVIS mode can be used successfully with very low-power HF sets. This will result in much lower probabilities of intercept/detection (LPI/LPD). Figures M-14 and M-15 show results obtained in Thailand jungles and mountains with the 15-watt AN/PRC-74 operating on one SSB voice frequency (3.6 MHz) over a 24-hour period.

15-watt AN/PRC-74 operating on one SSB voice frequency (3.6 MHz) over a 24-hour period.

M-8. Conditions Under Which to Use the NVIS Concept

Near-vertical incidence sky-wave techniques must be considered under the following conditions:

- The area of operations is not conducive to ground-wave HF communications (for example, mountains).

- Tactical deployment places stations in anticipated skip zones when using traditional frequency selection methods and operating procedures.

- When operating in heavy wet jungle (or other areas of high signal attenuation).

- When prominent terrain features are not under friendly control.

- When operating against enemy ground-wave jammers and direction finders.

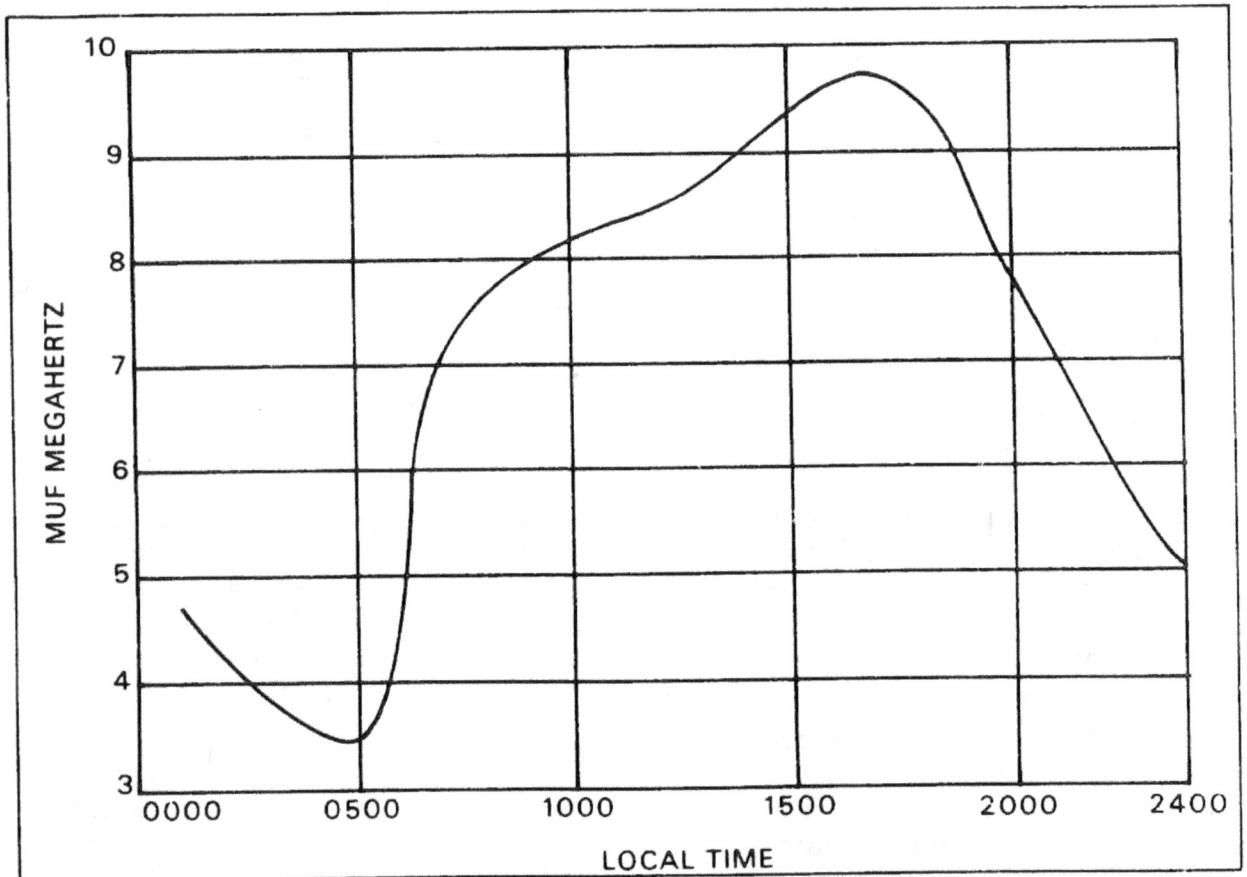

Figure M-13. Maximum usable frequencies in Vietnam.

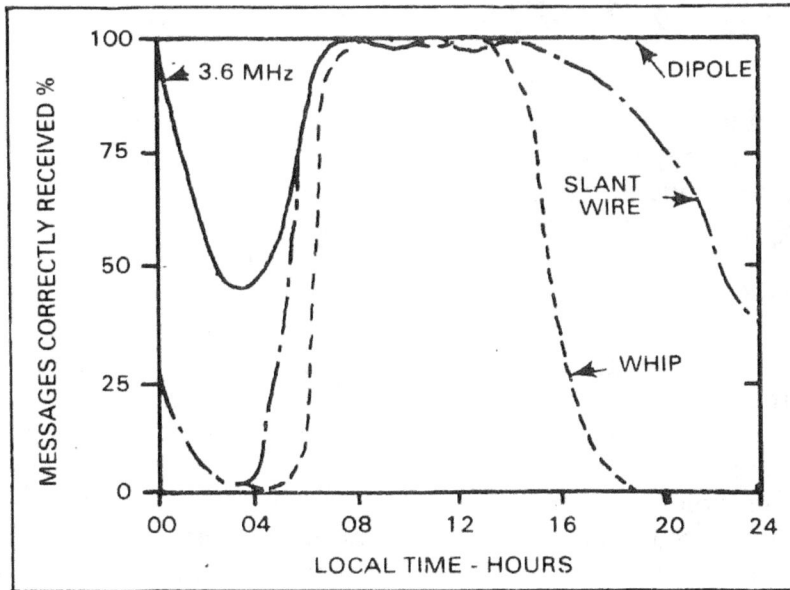

Figure M-14. Communications success with the AN/PRC-74 as a function of time of day and antenna type over a 12-mile path in low mountains, spring and summer 1963.

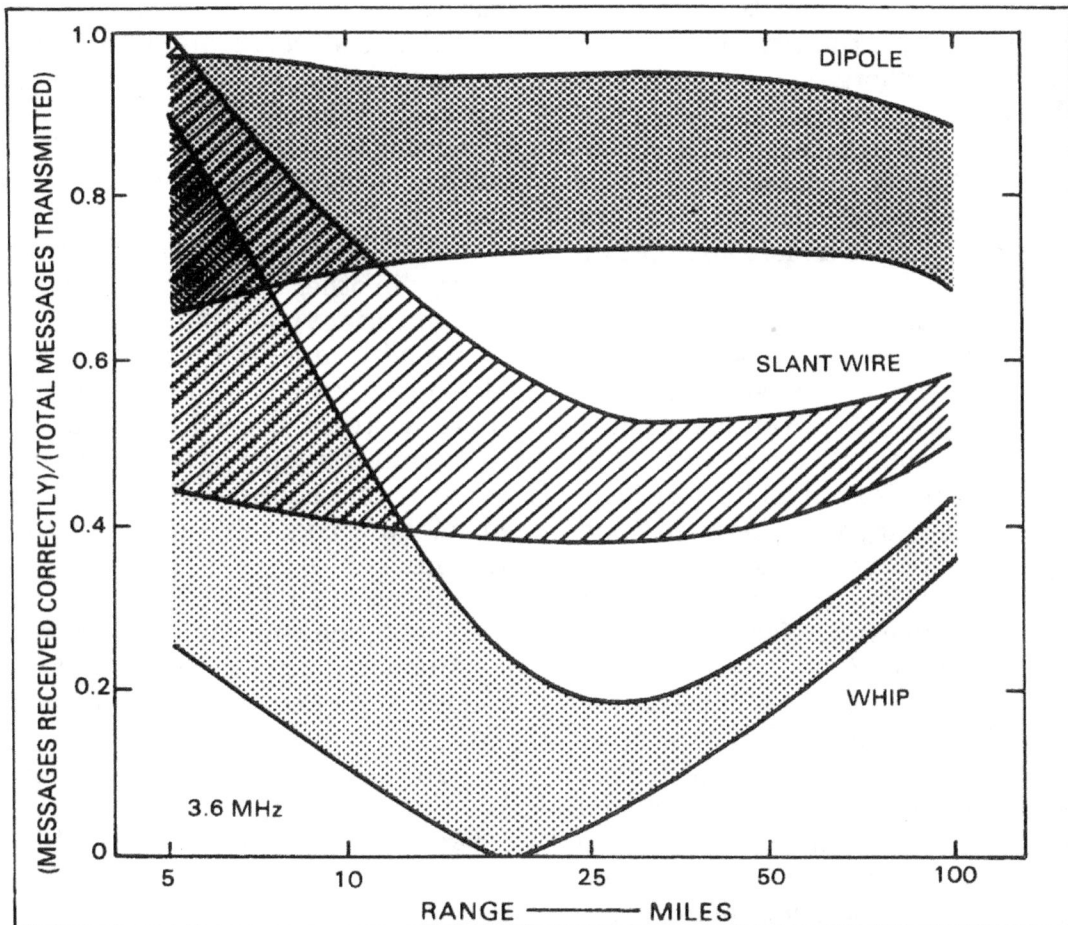

Figure M-15. Communications success as a function of range for the AN/PRC-74 in mountainous and varied terrain—including jungle in Thailand.

13: A NVIS REFRESHER

by Stanly Harter, KH6GBX

Scheduled high frequency radio nets over large areas with only one frequency should be recognized as being generally unrealistic. Propagation may dictate that more than one frequency is necessary. Conditions over which the Net Control Station may have no control often can ruin a net. This is particularly true on Amateur Radio bands.

The main reason a statewide HF net is unrealistic over an area as large as California, for example, is that the typical emergency involves only one location or area. Thus, the best frequency dictated by propagation characteristics is selected. The requirement is to communicate between Point A and Point B — not the entire state.

This is why station operators shouldn't be too upset about poor conditions between other stations and excellent results between others. All too often it is simply the laws of marginal propagation being in charge.

If wide area nets are really necessary to disseminate information or assure total station participation, sub-nets are necessary. This means breaking the larger area into perhaps two areas and changing to a more appropriate frequency to do so.

Antennas for DX serve no purpose in our nets. Near Vertical Incidence Skywave — or NVIS — antennas will improve your nets more than any other step. The NVIS antenna is just a few feet off the ground. A fixed station NVIS antenna is always horizontal and is installed, for practical purposes, anywhere between 7 and 25 feet (no more) above the ground. The same is true of a mobile HF antenna; it is always horizontal and never vertical. Now, doesn't that make your garage happy?

The horizontal antenna is a dipole cut to the lowest operating frequency. If it must operate on more than one frequency (and I don't know a service that can), it must either be a broadband dipole designed for this service (B&W is notable) or a single wire dipole connected to an external automatic antenna tuner (SGC, Motorola and perhaps others). Antenna tuners built into the HF transceivers do not qualify to do the job. The feedline required is

beyond the scope of this brief paper.

This summary is based upon the assumption that the reader has some familiarity with the subject of Near Vertical Incidence Skywave HF-SSB propagation for communications between 1.5 MHz and around 10 MHz. NVIS is essential to anyone requiring reliable HF communication from one to 400 miles. Such users include the RACES, Operation SECURE, the Civil Air Patrol, FEMA, the U.S. Forest Service, MARS and others. This information is not found in conventional technical publications and least of all in the field of Amateur Radio. Having said that, here again are the highlights of NVIS (pronounced "niviss").

For practical communications plans and operations, NVIS functions between 1.8 MHz to 10 MHz. Much above that and the signal penetrate the ionospheric layer instead of the desired reflection back to earth.

Using a "NVIS antenna" provides total coverage for a radius of 300 to 400 miles from any such station.

A NVIS antenna is always horizontal. A vertical antenna can never be used, including mobiles.

A NVIS antenna has omnidirectional radiation; in other words, it makes no difference how you orient your antenna.

A NVIS antenna is low; it MUST be low. Attempt to keep it no more than twenty feet above electrical or earth ground.

A multi-frequency NVIS antenna requires a remote and automatic antenna tuner at the end of the coaxial cable and before the antenna system.

If you use a dipole antenna with an automatic antenna tuner to operate on more than one frequency, cut the dipole to the lowest frequency to be used with the conventional formula.

An existing dipole antenna over twenty feet high can be expediently modified to obtain a degree of NVIS performance. This is done by allowing the feedpoint to stay ten to fifteen feet below the ends of the antenna.

A horizontal broadband antenna may be used without an automatic antenna tuner for a base station.

End-fed longwire antennas are NOT recommended. Unbalanced antennas are prone to creating interference to telephones and other electronic systems in the vicinity.

When the user has a choice of several frequencies, the best choice is generally ten percent below the MUF or Maximum Usable Frequency obtained from propagation programs.

Now, and for the next several years, solar activity will affect HF communicators in a manner to which most are not accustomed. Amateur Radio operators will find that 80 Meters will often work better in the daytime than 40 Meters and that 160 will be better than 80 at night.

The Civil Air Patrol will find that it must use 2347 kHz at night instead of 4585 for more reliable communications. Remember, we are talking about communications necessary up to about 400 miles away. This may mean some necessary equipment and antenna changes. One thing is certain — this condition will be with us for several years and justify the expense.

In summation, HF-SSB and NVIS will climb out of the deepest canyons, hop the highest mountains, never require any repeaters or other intermediate relays. And some thought high frequency radio was an ancient art!

PART THREE

YES, IT WORKS!

Acceptance for any new idea is usually slow and marked by caution. This is particularly true in military doctrine since the stakes are so high. NVIS techniques have been demonstrated since World War II and most recently thoroughly evaluated by the U.S. Marine Corps. Yes, it does work.

NVIS a key to D-Day success

by David M. Fiedler (LTC, NJANG)

The lesson of D-Day for the Signal Corps should be that HF radio and correctly applied antenna and frequency assignment techniques will provide the first in, last out, most reliable communications system possible to support combat operations.

Over the past several years, I have attempted through the pages of the ARMY COMMUNICATOR and by rewriting the applicable portions of FM 24-18 to impress our present day Signal Corps commanders and systems engineers with the combat capabilities of High Frequency radio systems, particularly those using Near Vertical Incidence Skywave (NVIS) high radiation angle antenna techniques.

These techniques, although neglected by the Signal Corps for many years while we developed SATCOM and other systems, are still highly effective and reliable combat communications techniques that will survive under tactical combat conditions that would destroy most other types of systems. At the same time, HF radio and antenna systems will provide longer ranges and wider areas of coverage at less cost than almost any other military communication means.

Until recently, it was not common knowledge among military communicators that NVIS HF radio systems played the key role in the success of US and Allied Forces during the D-Day invasion of Normandy in 1944.

NVIS techniques were incorporated into the Signal support planning for D-Day by Dr. H. H. Beverage, a well known early American radio pioneer who was serving at that time as a member of the United States Special Technical Advisory Force (USSTAF) Communications Advisory Specialist Group (CASG).

Just before D-Day, Beverage arrived in England and found that communications problems were causing grave concern on the part of senior allied ground and air commanders because poor radio communications were causing severe disruptions in the command and control of the tactical air support effort to the invasion asssault force. The senior commanders felt, and justly so, that

unless crushing, superior, air support was brought to bear upon the enemy defending Utah and Omaha Beaches, the invasion would be thrown back into the English Channel by the Nazis in the first hours of the assault. The commanders also felt that the key to gaining superior air support was good radio communications, which would provide the command and control edge necessary for victory.

Beverage went directly to the headquarters of the 9th Tactical Air Force (9th AF) at Uxbridge near London, which was to function as the air control headquarters for the assault force and controlled 9th Air Force (9th AF) air assets located all over England. Uxbridge was supposed to be in direct radio contact with the USS ANCON, a command and control ship located with the invasion fleet. An Admiral and an Army General aboard the ANCON controlled all air attack and naval gunfire support to the invasion force as well as to the troops in the assault force.

The operation plan called for a forward communications center on each of the landing beaches; the USS ANCON and 9th AF headquarters at Uxbridge were to operate on a single HF radio net with the shore stations sending air support requests and targets to Uxbridge through the ANCON. The 9th AF Commander would then reply via a coded radio messages to the USS ANCON stating the number of fighters and bombers assigned and their time of arrival. Officers aboard the USS ANCON would then feed the aircraft into the invasion area by assigning targets and times of attack to them, so the thousands of aircraft involved would not run into each other or be shot down by friendly forces. This system is shown in Figure 1.

During the preparation for the invasion, Beverage found that Uxbridge was experiencing severe problems in getting the message

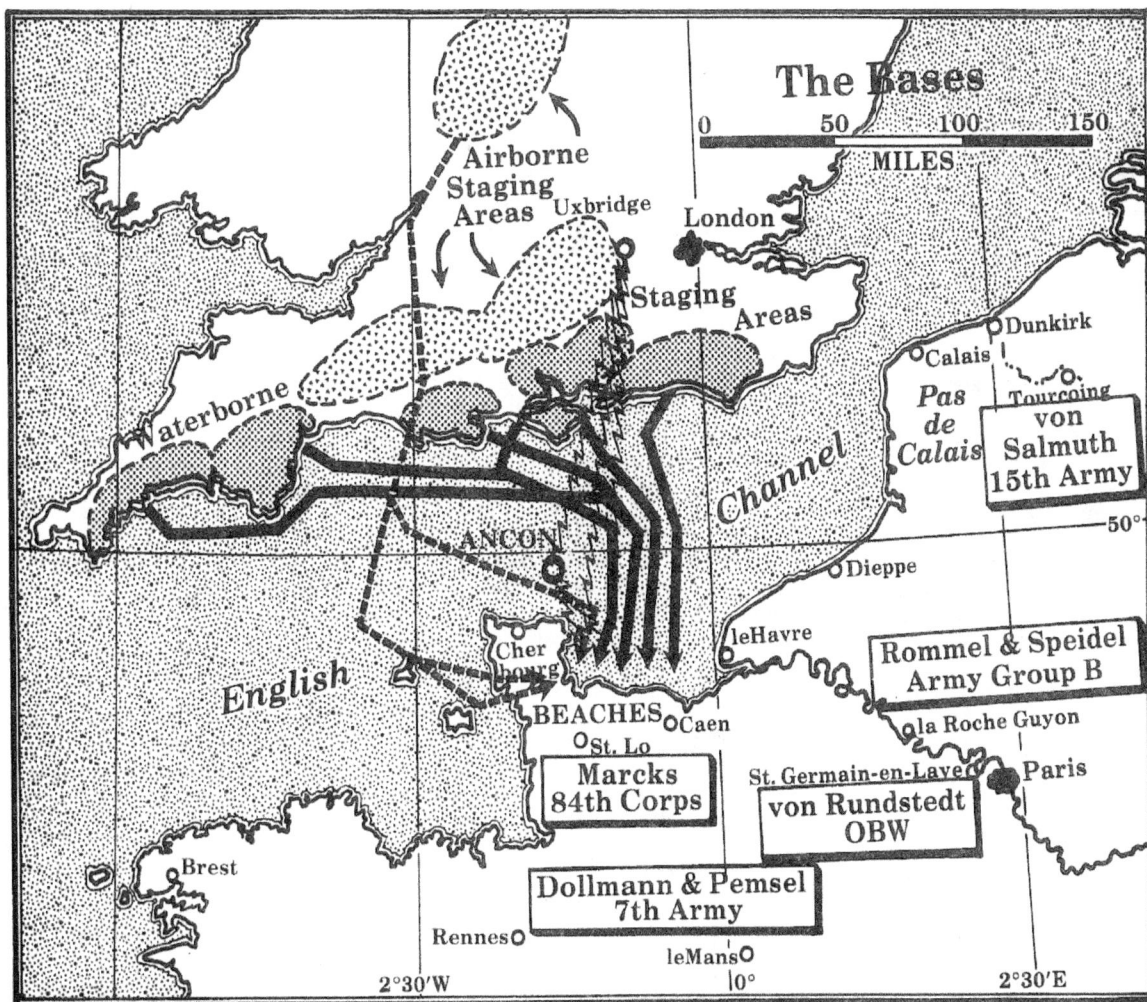

Figure 1.

traffic to and from the USS ANCON and expected that when the assault forces portable radios were landed on the invasion beaches, more problems would occur because of the lower power of the radios, expected antenna difficulties and all the other problems encountered when operating a tactical radio station in combat.

Beverage later stated, "We found that a vertically polarized antenna (whip antenna) was being used (at Uxbridge) similar to ones used at locations overlooking the English Channel. The propagation of vertical polarization over sea water is excellent; however, Uxbridge is about forty miles over land in the direction of the USS ANCON and Omaha Beach in France, and thus presents a different problem. The ground wave over land is very poor on both vertical and horizontal polarization. The only

solution is the use of skywave with horizontal polarization (NVIS). Many days just prior to the invasion were spent supervising the change to skywave (NVIS). I was worried, with the change whether or not the signals to and from the USS ANCON and the troops going ashore would get through. On D-Day I was briefly confused because all frequencies had been changed to confuse the Germans. I was greatly relieved of my anxiety when I soon knew that all had performed as planned."

Beverage, later received a letter from Gen. Brereton, Commander of 9th AF, which read "It is my desire to officially command Dr. H. H. Beverage of the USSTAF for service performed for the 9th AF during the period 20 May 1944 to 16 June 1944. Beverage furnished invaluable technical advice and assistance in the

establishment of 9th AF radio systems utilized in the initial stages of the invasion of the Continent. As a result of this work, NO technical failures in the cross channel radio circuits were experienced. Such an achievement would not have been possible without his help."

Such a lesson should not be lost on modern Signal soldiers and Marines. The D-Day invasion, which was the largest, most critical operation of its type ever conducted in the history of warfare, could not have succeeded without the necessary close air support provided by the greatest tactical air force ever assembled. This air support would not have been possible without NVIS communications.

As we all know, history repeats itself because it is trying to teach us a lesson. The lesson of D-Day for the

Signal Corps should be that HF radio and correctly applied antenna and frequency assignment techniques will provide the first in, last out, most reliable communications system possible to support combat operations.

Under the recent reorganization of the Signal Corps due to MSE, most HF radios have been removed from division and corps Signal battalions, and only half of the required Improved HF Radio (IHFR) procurement has been funded. HF radio equipment and tactical systems have become a combat user and not a Signal Corps responsibility. Untrained users of HF radio tend to use only the vertical whip antennas that come with their equipment and then stop when they are unsuccessful in operating over wide areas or long ranges. The present HF radio situation in the Army has many parallels to the poor situation that existed in Europe prior to the arrival of Beverage. Let's hope that we don't have to learn D-Day's lesson over again.

Signal officers, particularly those who are assigned as corps and division Signal officers and assistant corps and division Signal officers as well as Signal officers assigned to brigade and battalion staffs must be acutely aware of the application of HF radio and antenna techniques now that the vast majority of HF equipment resides outside of the Signal battalions and within the units of other (non-Signal) branches. It is the heavy responsibility of these officers to assure that the proper frequency assignments and antenna techniques are employed and that the non-Signal battalion personnel operating our HF radio equipment are trained properly. With the lessons of D-Day in mind, it would also be wise for the Signal Corps to reevaluate the wisdom of removing the HF radio equipment (primarily HF radio teletype) completely from the Signal battalions, thus relying strictly on SATCOM for Beyond-Line-of-Sight (BLOS) communications on the battlefield. In my opinion, it is extremely critical to add both a single channel SATCOM and an HF radio capability to the Combat Net Radio (CNR) interface facility in MSE. This capability will make MSE capable of beyond-line-of-sight communications quickly and cheaply, since all required equipment already exists including the MSE facility into which the equipment must be installed. This improvement will then make MSE more capable of supporting the AirLand Battle concept expressed in our doctrine.

It is also time for the Signal Corps and the Army to re-examine the wisdom of removing HF radio teletype from the Signal battalions. This action has severely reduced the Army's beyond-line-of-sight tactical communications capability and its ability to fight AirLand Battle. It has also removed one important means presently in extensive use for interoperation with allied forces. We must, in my opinion, retain a minimum number of HF radio teletype assemblies in each division and corps Signal battalion sufficient to service the major elements of the divisions and corps when they are located beyond-line-of-sight and also to serve the allied interface function. To do this, the existing facilities should be reequipped with modern reliable HF radios, and MODEMS capable of error detection and correction. In addition, meteor scatter and SATCOM capabilities should be added to the facility as well as an MSE interface so that a "triple threat" capability exists for accomplishment of the beyond-line-of-sight communications mission.

When this is accomplished, then we can truly say, that the Signal Corps has learned the lessons of D-Day.

References

Genius at Riverhead, a Profile of Harold H. Beverage *by Alberta I. Wallen.*

The Longest Day, June 6th, 1944 *by Cornelius Ryan.*

State of the Signal Regiment Information Papers, *17th Annual Signal Conference, December 5-8, 1989.*

Mr. (LTC) Fiedler is a graduate of SOBC, SOAC, the Radio and Microwave Systems Engineering Course, and the Command and General Staff College. He has served in Regular Army and National Guard Signal, Infantry, and Armor units in both CONUS and Vietnam. He holds degrees in both physics and engineering, and an advanced degree in industrial management.

15: MARINE TESTS PROVE FIEDLER'S NVIS CONCLUSIONS

Lt. Col. David M. Fiedler, NJARNG

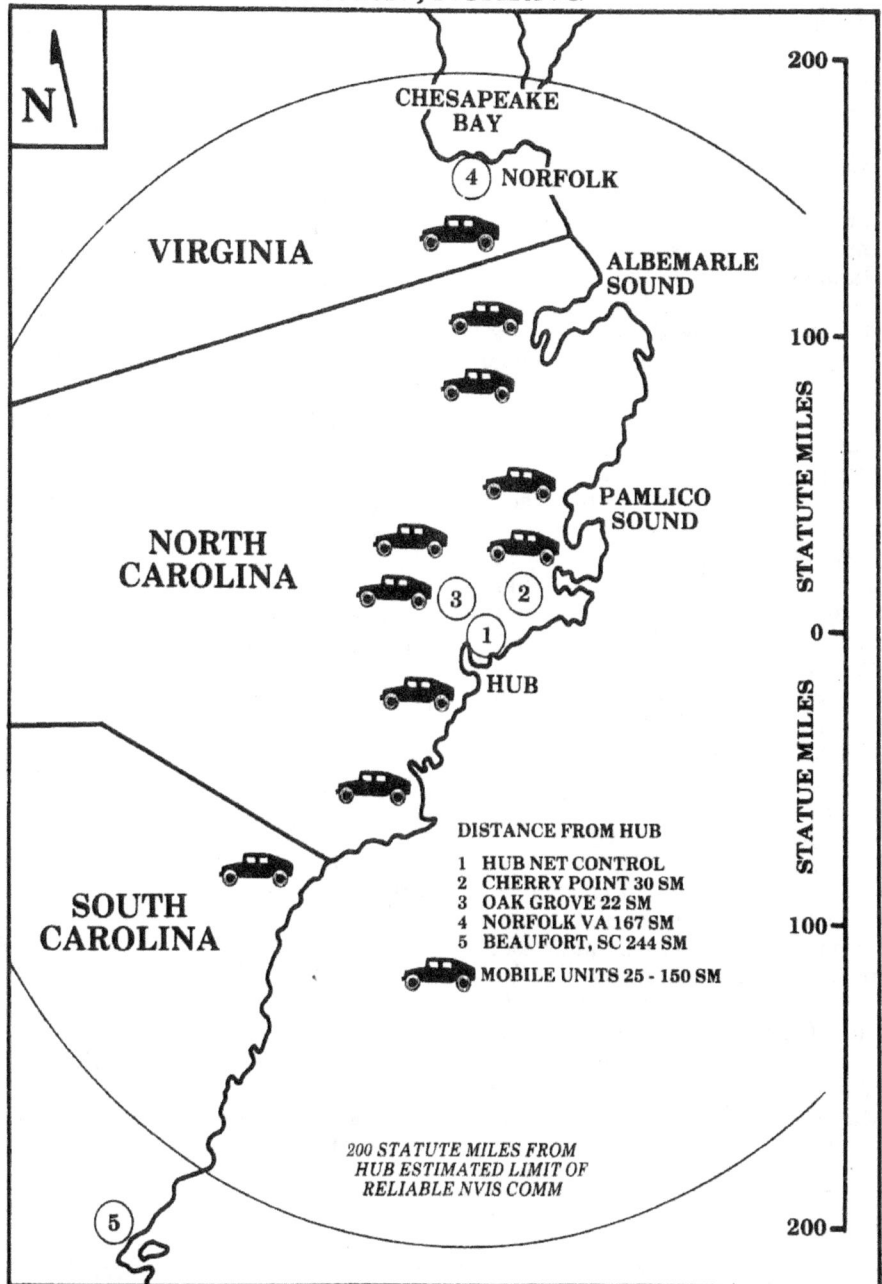

Figure 1. Area of test operations

There is a place in our architecture for all means of transmission, and to remove a proven, quickly implaced, highly reliable, and highly survivable means of tactical communications instead of improving it is a big mistake, which must be stopped.

For many years, our colleagues in the U.S. Marine Corps (USMC) have faced communication problems identical to Army and Army National Guard communicators trying to communicate over corps size areas without gaps in communications coverage. In Marine Corps operations, the transmission ranges required to communicate between the landing beaches and the forward line of troops often can exceed 100 nautical miles with critical units spread throughout this entire area.

Because of this Beyond Line of Sight (BLOS) requirement and a decided lack of SATCOM assets, my articles about NVIS HF Skywave Beyond Line of Sight Communications published in the ARMY COMMUNICATOR (AC) Fall 1983, Spring 1986, Winter/Spring 1987, Fall 1987, and Spring 1989, sparked a great deal of interest among USMC communicators.

In the Spring and Summer of 1988, units of the Fleet Marine Force Atlantic (FMF Lant) conducted two separate large scale communications exercises in order to evaluate my conclusions about the suitability of NVIS techniques for employment in beach heads, landing zones and other USMC areas of operations using standard HF radios with fixed and mobile antennas specifically designed to achieve NVIS effects.

The results of this effort were published in the **Marine Corps Gazette** in March 1989, and in a Marine Corps test report in June 1989. The test was conducted over the area shown in Figure 1 in a hub spoke configuration by personnel of the 8th (USMC) Communications Battalion under the direction of Lt. Col. Tilden V. Click (USMC) and 1st Lt. Mark W. McCadden (USMC). The hub of the configuration was at Camp Lejeune, NC, with out stations at Cherry Point, NC (30 miles), Oak Grove, NC (22 miles), Norfolk, VA (167 miles), and Beaufort, SC (244 miles). Two mobile stations consisting of standard AN/GRC-193 radio sets mounted on HMMWVs (AN/MRC-138s) were moved through the test area at all ranges and azimuths in order to confirm my assertation (see AC Spring 1986) that no skip zone existed when using NVIS techniques. AN/PRC-104 20 watt manpack radios were also used to confirm operational findings using low power (see AC Spring 1989). All fixed stations used standard (AN/GRA-50 type) half wave dipole antennas at heights approaching .25 wavelength at the operating frequencies. Frequency assignments were made carefully in order to assure that the radio signals would be reflected by the ionosphere, since signals with frequencies higher than the so called critical frequency penetrate the ionosphere and are not returned to the earth (see AC Fall 1983).

Mobile stations used the standard 32 foot whip antenna (AT-1011) bent 90 degrees by using a whip tilt adaptor, designed by Allan Christianson (USAF ret.) now of ASC, LTD, O'Fallon, Illinois, and recommended in my AC Spring 1989 paper (see Figure 2). A second option was a 32 foot wire to simulate the AT-1011. Circuit reliability was calculated based on a voice communications check made every 15 minutes during the test period (48 hours). The results were:

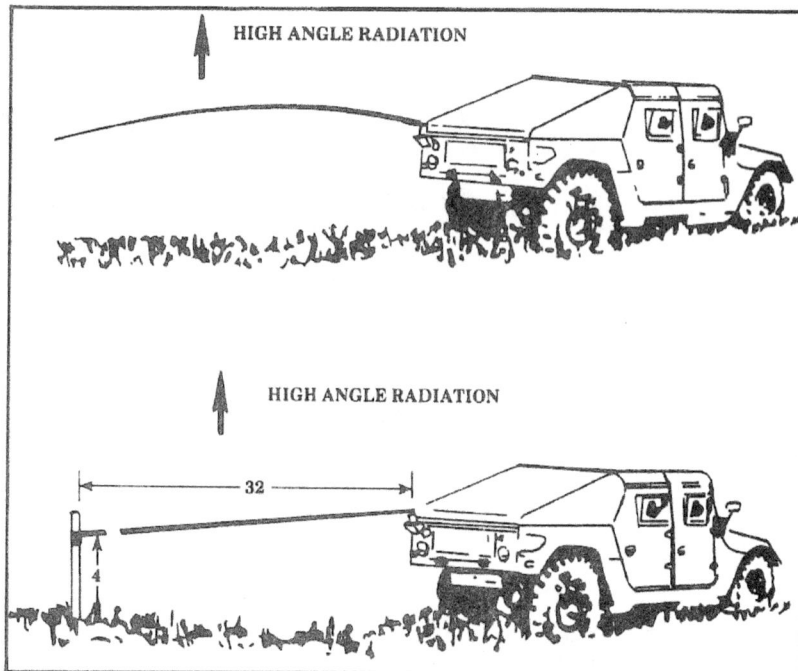

Figure 2. Mobile antenna configurations using AN/MRC-138

Two mobile stations (see Figure 2) traveled north and south of Camp Lejeune and stopped every 25 miles. Using either the 32 foot bent whip or the 32 foot wire produced identical circuit reliability results to the fixed station test over the many stops and checks that were conducted in the operational area. The results were:

104 low power (20 watts) manpack radios with standard 10 foot vertical whip antennas were employed at Camp Lejeune. Attempts to contact the other stations including Oak Grove only 22 miles away failed since the 10 foot whip radiated no high angle skywave energy, and the ground wave (low angle) signal

Mobile results				
	Camp Lejeune	Cherry Point	Oak Grove	Norfolk
Mobile 1	100%	100%	100%	100%
Mobile 2	100%	100%	100%	100%

Note, while 100% circuit reliability was attained using both the bent AT-1011 and the 32 foot wire (see Figure 2) operators reported that the 32 foot wire seemed to have a better audio quality. The reason for this was not assertained since both methods were more than acceptable (100% useable) in terms of operation, and no test equipment was available. The losses caused by the use of a sectional (screw connected) conductor may have contributed to this effect.

After the fixed and mobile tests were completed, a test was conducted to confirm that the outstanding test results were indeed the result of NVIS propagation. For this test, AN/PRC-

radiated by the 10 foot whip was attenuated by the path losses to the point where it was not sufficient for communication at Oak Grove. Once it was established that the AN/PRC-104 with the vertical whip could not communicate in the net, it was then connected to the 32 foot horizontal wire antenna in order to produce NVIS radiation. The result? Not only was communications established with Oak Grove, it was established with all other stations in the net as well. This proved dramatically, that antenna and frequency rather than transmitter power are the deciding

Fixed stations				
	Oak Grove	Cherry Point	Norfolk	Beaufort
Camp Lejeune	100%	100%	100%	67%*

* No satisfactory explanation for this lower rate was ever made during the test. Examination of the data from the fixed and mobile tests tend to indicate that equipment failure or operator error was most probably the cause, but no specific cause was ever confirmed, so this data should be discarded for both fixed and mobile results, in view of the other test data gathered.

Inverted L antenna.

Inverted L antenna, vertical pattern,
height 40 feet, length 150 feet.

——— 3MHz
- - - 4MHz

dBi

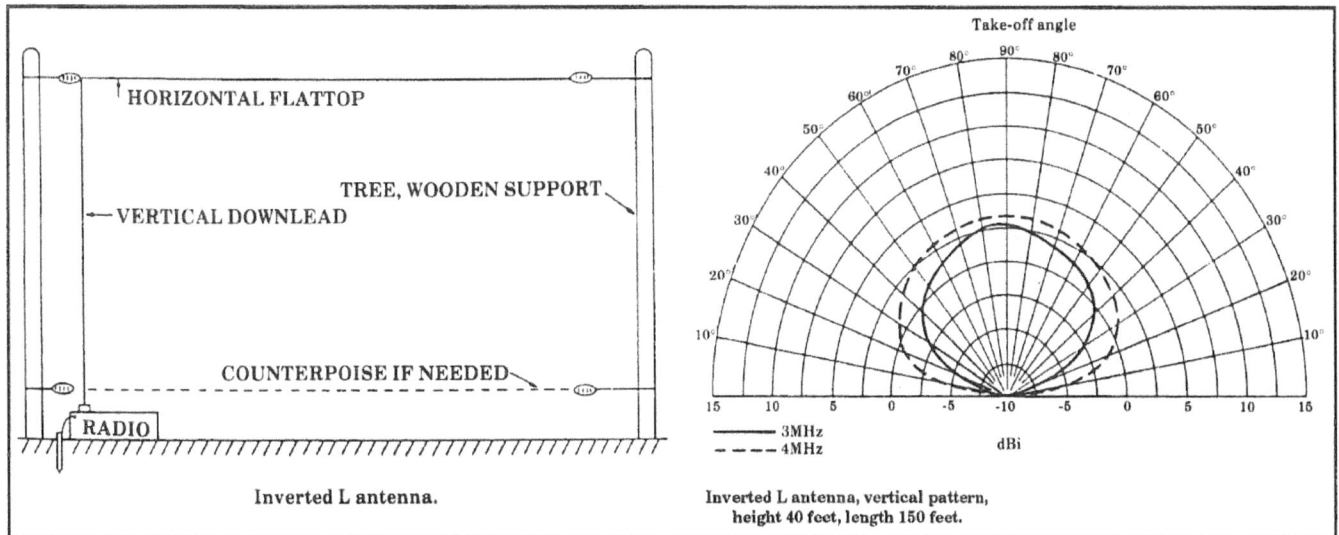

Figure 3. Inverted "L" antenna

Fixed stations tests "Inverted L"				
Camp Lejeune	Cherry Point 100%	Seymore Johnson 100%	Ft. Bragg 100%	Myrtle Beach 100%

factors when communicating via skywave on HF radio. Indeed, a radi of one twentieth the power of the AN/GRC-193 produced identical operational results for this test.

As a check against the results of this test which confirmed tests previously conducted by troops of the New Jersey National Guard (see AC), the Marines of the Communications Company, Headquarters and Service Battalion, 2nd Force Service Support Group (FSSG) FMF Lant under the Command of J. C. Fox (USMC) and 2nd Lt. M. S. Flannery (USMCR) conducted similar testing. In this test, Camp Lejeune was net control with Cherry Point, NC (30 miles), Ft. Bragg, NC (105 miles), Myrtle Beach AFB (116 miles), and Seymour Johnson AFB (57 miles) serving as fixed stations. All fixed stations used "inverted L" antennas 40 feet high and 100 feet long (see Figure 3) to show that simple wire antennas— such as dipole, "inverted V", "inverted L",—with good high angle radiation characteristics (NVIS) are ideal for HF radio communication over modern division and corps size areas. All of these antennas can be fashioned from the standard AN/GRA-50 antenna kit or any ordinary wire and insulators as required. The "inverted L" fixed station results over the test period (72 hours) were:

In this test, the 32 foot vertical whip was connected to the radios (AN/GRC-193s) and placed into the vertical position, and no communications was achieved thus providing no ground wave path existed between locations, and the above results were caused by NVIS propagation.

Once fixed station communications were established, the AN/MRC-138 with whip tilt adaptor (provided by AS Christianson Associates) was dispatched into the area of operations. The whip was shortened to 1/2 the 32 foot length of the antenna of the first test (16 feet) so that a more handy configuration and the effect of a shorter but still horizontal antenna could be observed.

The mobile station then traveled throughout the area bounded by the fixed stations and made radio contact with each fixed station approximately every six minutes for the entire day. In order to test the effects on the ionosphere of the time of day, night tests were conducted after proper frequency adjustments had been

made by driving the mobile station to Roseboro, NC, and back after dark and performing communication checks on the same six minute schedule.

In order to test the effects of antenna and vehicle orientation, a route was selected that would cause the mobile station to travel in a Figure 8 pattern thus presenting every possible azimuth to every fixed station in the test. This single drive took over six hours to complete.

The result of all this driving? A 90% probability of communication between the fixed station and the mobile unit with the 16 foot horizontal whip antenna and AN/GRC-193 radio. Very impressive. These results are all the more impressive when one considers the high level of atmospheric noise that existed within the test area during the tests. Heavy thunderstorms with considerable lightning were present during the test period. Despite the lightning, the inefficient 16 foot antenna and the inherent high noise level of the frequency band being used 90% reliability for voice communications was still achieved. These results far out-perform VHF line of sight combat net radio results when operating in this type area. The ranges and areas covered were many times greater than VHF-radio could provide, thus proving that HF NVIS can be highly valuable on the modern battlefield and is a viable method of beyond line of sight communications.

What have we learned from the USMC tests? I think we have proven the following in the field which is where it counts:

• Antenna and frequency are the paramount design and operational considerations when employing combat net radios on the battlefield (true not only for HF but all radio communications).

• Current family of HF radio equipment. AN/GRC-193, AN/PRC-104) are more than adequate to cover division and corps size areas if proper antennas and HF frequencies are selected, and the equipment is used properly.

• Simple wire antennas that can be fashioned cheaply from the standard AN/GRA-50 antenna kits or plain wire are sufficient for division and corps operations.

• While at brief halts or on the move, highly reliable communications to mobile stations can be achieved by simply bending the standard vehicle whip antenna into a horizontal position. This can be done with whips as short as 16 feet; however, longer is better. To do this, a whip tilt adaptor must be brought into the Army inventory.

• No "skip zone" or gap in communication coverage exists when using NVIS techniques.

To those in the Signal Corps who have advocated the removal of HF radio and HF radio teletype (RATT) from the Signal Battalion (a course on which we are now embarked) in favor of other forms of battlefield tactical communications and to those who would saddle the tactical combat forces with intricate high cost antenna arrays requiring considerable areas for antenna deployment, I say the Marines have proved that our current modern standard HF radio and antenna systems can cover greater areas, more reliability, with less cost under almost all conditions, when compared to these other means of communications if you know what you are doing. I say that there is a place in our architecture for all means of transmission, and to remove a proven, quickly implaced, highly reliable, and highly survivable means of tactical communications instead of improving it is a big mistake, which must be stopped.

To those who do not agree, Teddy Roosevelt said it best: "Tell it to the Marines."

I wish to thank the officers named in this article as well as 1st Lt. Roger W. Roland (USMC) for having faith that what I was telling them was true and proving it. I also wish to thank Allan Christianson

(USAF ret.) for providing valuable technical information and prototype whip tilt adaptors at a great price (gratis).

References
*NVIS: "Highly Successful Rarely Used",
U.S. Marine Corps Gazette, March 1989*
Test Report, "Near Vertical Incident Skywave (NVIS) Mobile Communications Exercise (1500 10/OPS/R 23 Jun 89)", Communications Company, 2nd Force Service Support Group, Fleet Marine Force Atlantic, Camp Lejeune, NC 28542

Mr. (Lt. Col.) Fiedler was commissioned in the Signal Corps upon graduation from the Pennsylvania Military College (Weidner University) in 1968. He is a graduate of the Signal Officers Basic Course, the Radio and Microwave Systems Engineering Course, the Signal Officers Advanced Course, and the Command and General Staff College. He has served in Regular Army and National Guard Signal, Infantry, and Armor units in both CONUS and Vietnam. He holds degrees in both physics and engineering and an advanced degree in industrial management.

He is presently employed as the Chief of the Fort Monmouth Field Office of the Joint Tactical Fusion Program (JTFP), and is the Assistant Program Manager (APM) for Intelligence Digital Message Terminals (IDMT). He is also the Director of Systems Integration for the JTFP. Concurrently, he is the Chief of the C-E Division of the NJ State Area Command (STARC), NJARNG. Before coming to the JTFP, Fiedler served as an engineer with the Army Avionics, EW, and CSTA Laboratories, the Communications Systems Agency (CSA), PM-MSE and PM-SINCGARS. Fiedler has published several important articles on tactical communications and electronic warfare.

16: A CLOUD WARMER ANTENNA IS BEST FOR LOCAL HF COVERAGE

by Stanly E. Harter, KH6GBX

There has been what can be called more than somewhat mild excitement in Northern California emergency communications circles over a form of high frequency radio propagation. It's not new, but I venture to say that very few have used and understood it. "It" is called NVIS — Near Vertical Incidence Skywave.

Patricia Gibbons, WA6UBE, presented a paper on NVIS at the 1990 Pacific Division ARRL Convention in San Jose. It caused quite a stir. She quickly ran out of handouts and has since received dozens of requests for more. The handouts included reprints of articles from military communications magazines reporting the results of many tests.

Near vertical incidence skywave means forcing your radio signals to travel straight up (i.e., 80-90 degrees) and back down. This achieves radio coverage in circle having a radius of 300 miles and more. Stop and think about that for a moment. Complete coverage within such a circle on frequencies between 2 and 10 or 12 Megahertz.

Some readers may wonder what's so good about this. So now is a good spot to say that if only DX (long distance) is your thing, skip on and read one of the other fine articles in this publication. We are talking about dependable local area high frequency communications — the type we need for tactical public safety communications in the Radio Amateur Civil Emergency Service, the Civil Air Patrol, SECURE, search and rescue, forestry, pipeline and similar services. In tactical communications we don't want DX.

How frustrating it was in years gone by to drive away from, say, a 4585 kHz base station, only to lose a good 400-watt signal a mile from the transmitter! All the while receiving, loud and clear, a 50-watt transmitter some 200 miles away. Very frustrating. We really didn't know why. When VHF-FM radios and repeaters came along, most of us retired HF mobile radios for tactical communications.

The reasons we haven't enjoyed good HF tactical communications, whether AM or SSB, have been the base and mobile antennas. The classic dipoles, a quarter- to a half-wave up in the air. The mobile antennas, designed for use by Amateur Radio operators, have the same general propagation characteristics — low take off angle for DX.

Virtually every Amateur Radio mobile HF antenna is unsuitable for day-to-day tactical communications. They are variously bulky, mechanically weak, won't survive continual whacks from limbs and low overhead, look like Neptune's trident or a misshapen coat rack. They may be fine for hobby communications but not for tactical public safety use. In that type of service we want one, simple antenna that is permanently installed and we don't have to think about or fuss with again.

So how do we achieve NVIS? By getting those sky hooks down near the ground. Let's start with the base station antenna. Horizontal, of course. Dipole or long wire. Place the antenna as low as two feet above the ground but no higher than about thirty feet without a counterpoise. Use an appropriate and sturdy antenna tuner; you will use the one antenna for all frequencies between 2-12 MHz.

A longwire antenna is suitable in field setups but not recommended on office buildings or other urban environments. The reason is that unbalanced antennas frequently create interference problems with telephones and other communications and electronic equipment. These problems are substantially reduced or eliminated with a balanced antenna system.

The antenna tuner of preference is one that is automatic. Such tuners are available now that do not require any control cables; they require only the coaxial transmission line from the transceiver and a 12-volt DC cable. The tuner is placed at the far end of the coaxial cable. There are then two basic options: A longwire or a balanced (dipole) antenna. The longwire can be any length — the longer the better, to approach the lowest operating frequency. A very good ground connection is necessary and often quite difficult to obtain on a rooftop. (When we are talking about running ground connections we mean the shortest possible runs of 2- to 3-inch copper strap — never wire or braid.) For a balanced antenna, you can place a 4:1 balun on the output of the antenna tuner, thence to a 450-ohm feedline to the dipole antenna. Any NVIS antenna can be enhanced with a ground along the surface that is 5% longer than the antenna and separated by .15 wavelength at the lowest frequency to be used.

For the HF-SSB mobile radio, a sixteen-foot whip is probably the best. Such a whip may be both costly and difficult to find. For NVIS, the antenna is used folded down, both in motion and at rest. That's right, it is not released to go vertical. Most us use the heavy duty ball joint mount, heavy duty spring, and readily available 106-inch whip.

To further improve the NVIS propagation at rest, the mobile whip is adjusted to go parallel to the ground and away from the vehicle. A further enhancement is to remove the whip and run out a longwire 30, 50, 100 feet long. Patricia Gibbons carries orange traffic cones, about 18 inches tall, and notched at the top to lay the antenna wire away from the vehicle.

The Russian military have been using NVIS antennas on their vehicles for quite some time. They appear to be about 4 meters long and about six inches above the top surface of the vehicle. At least one American manufacturer makes a NVIS antenna for both military and civilian vehicles. On a van it looks no more obtrusive than a luggage rack.

The automatic antenna tuner is located in the rear of the vehicle and as close as possible to the mobile antenna feedpoint.

An HF-SSB mobile radio was recently installed in one of our State Office of Emergency Services trucks. The installer and the vehicle were 80 airline miles away and the time was 2 p.m. In the State SECURE (State Emergency Capability Using Radio Effectively) system this calls for using a 7 MHz channel. We established contact; the mobile signal was received here in Sacramento at about S5 to S6. I then asked him to loosen the ballmount, flop the antenna down horizontal and away from the truck. I could tell by the pause and tone of his voice that he thought I had lost it. When he returned

to the air his signal jumped to S9. By the same token he thought I had cut in a linear amplifier because of the improvement to my signal. I assured him that the improvement was due solely to his flopping his antenna horizontal.

You need not be concerned over the orientation of a NVIS antenna; it is omnidirectional.

Every Monday night from 7-8 p.m. we conduct a State RACES net on 3545.5 kHz using AMTOR. One night the net was concluded and secured. While the hams were cleaning up one of them noticed that we were being called; there it was on the screen. But it belied the loudspeaker; there were no discernible AMTOR signals — only a high noise level. Yet, there was that station, WA6UBE, calling us at W6HIR. Yes, it was Patricia Gibbons proving a NVIS point again. She was transmitting to us from 82 miles away with an antenna lying on the ground along her driveway and using 3 watts of power!

On another statewide evening RACES net, our Monday night 8 p.m. 3952 kHz voice net, Bill Pennington, WA6SLA, compared two antennas. One was a vertical and the other was a horizontal quite close to the ground. His observations were interesting and typical of NVIS propagation. Almost all of the signals received on the vertical were higher in voltage than the NVIS antenna but, be that as it may, *the signal-to-noise ratio is superior with the NVIS antenna*. The noise floor is measurably lower on the lower antenna, thereby providing better overall communications.

I heard more than one amateur say, after listening to Gibbons' NVIS presentation and subsequent demonstrations, they decided to jump back into HF-SSB mobile radio again. These people, an d I also, are interested primarily in the mobile tactical public safety communication applications.

There is an easy method to improve the NVIS radiation of your dipole antenna. Let the feedpoint sag five to ten feet below the horizontal. This will alter the radiation to improve the vertical angle to achieve an approximate 2 dB improvement at no cost.

Many are excited about an old, but little understood and practiced, means of HF radiation. If you need it, try it. You'll like it.

17: BATTLE FORCE ELECTRONIC MAIL

An Electronic Mail System for Half-Duplex Radio Link Applications

by Terry A. Danielson

Naval Command, Control and Ocean
Surveillance Center
RDT&E Division, Code 846
San Diego, CA 92152

Editor's notes on "Battle Force Electronic Mail"

High frequency radio is commonly used for intership communications. The usual propagation mode among ships deployed over the area typically occupied by a carrier battle group would probably be surface wave, not NVIS.

Mr. Danielson's group has also deployed the system described on a research submarine. It was able to maintain contact with its base in San Diego, including Internet access, over distances exceeding a thousand miles. This indicates some HF skywave propagation mode was involved.

Of course how propagation occurs isn't very important to the system Mr. Danielson describes. The editors have included it because it illustrates the versatility and value of HF radio when used in an insightful and well executed application.

HF nodes are relatively easy to deploy, even under emergency conditions. Care with propagation and antennas results in links capable of maintaining digital communications at 4800 bps over NVIS paths. Message throughput and accuracy is dramatically improved as compared to combat net radio or voice message handling techniques.

The technique described provides communications security by means of conventional KG-84C military encryption equipment. With proper message handling discipline it is possible to use civilian encryption techniques (e.g., Clipper or DES, the U.S. Federal Data Encryption Standard). Properly implemented, extremely secure communications are possible. This is possible because of the digital nature of the messaging — comparable security is impossible with analog voice techniques. (Note: The Amateur Radio Service is prohibited from using encryption.)

The Navy has provided a very interesting and thought-provoking example of the modern effective use of HF radio.

•

When the ships of the ABRAHAM LINCOLN Battle Group deployed in April 1995, they departed with a new capability for secure, error-free exchange of electronic mail between ships Unlike most new communications capabilities provided to U.S. naval vessels in recent years, this one uses high frequency radio, hence its initial name. HF e-mail. Until the LINCOLN Battle Group (BG) deployed with HF e-mail, intership communications were limited to teletype nets and voice circuits — none of which could support the computer-to-computer exchanges necessary for conducting business in the information age. HF e-mail links the battle group with Local Area Network (LAN)-like connectivity, even across several hundred miles of ocean.

HF e-mail was born out of the Chief of Naval Operations (CNO N6) and the Space and Naval Warfare Systems Command's (SPAWAR PMW-172) interest in providing better intership connectivity at minimal cost. They, in turn, tasked the Naval Command, Control and Ocean Surveillance Center, RDT&E Division (NRaD), the Navy's C41 Laboratory in San Diego, CA, to see if HF radio could be used to accomplish the task. E-mail was chosen as the transmission medium because it is the only non-satellite medium capable of meeting the connectivity requirements of dispersed Carrier Battle Group or Amphibious Ready Group operations. A basic system block diagram is shown in Figure 1.

The HF e-mail system provides secure, error-free automatic delivery of e-mail messages, and binary files such as images and graphics. The system is built around the Amateur Radio "JNOS" network operating software which was modified by NRaD to operate at higher speeds and to accommodate military crypto devices such as the KG-84C and the KIV-7. JNOS is a highly capable and flexible networking software that uses standard Internetworking protocols and packet transmission techniques. This MS DOS-based software is capable of supporting networks of several ships, and can serve point-to-point links as well. The "HF e-mail" system has been used operationally over UHF radio transceiver links in areas in the Middle East where surface ducting provides long range connectivity using line-of-sight media, and has been successfully tested over UHF DAMA satellite and VHF SINCGARS radio system as well.

The system half-duplex protocol interface to the HF modem is via an RS-232 synchronous serial communications port. The crypto, if used, is inserted between this interface and the modem. The HF transmission modem selected is one of the commercially available products which provide the single-tone serial waveform per MIL-STD-188-110A. This is presently the most robust high speed waveform for the HF channel, and can support speeds up to 4.8

kbps. However, modem performance becomes less robust as the user data rate is increased, thus the system works best for speeds of 1.2 to 2.4 kbps. At least one vendor is working on a waveform capable of supporting 9.6 kbps over ground wave channels where multipath (a skywave phenomenon) is not a factor. The system has also operated successfully with the NATO STANAG 4285 waveform.

The modem audio and keyline interface goes directly to the radio equipment. This connection is made aboard ship via patch panels and switchboards For tactical radios such as manpack transceivers, entry is via the handset connector. The system works over conventional single sideband (SSB) HF radios, with the only requirement being reasonable phase stability and a nominal 3.0 kHz bandwidth. Narrower bandwidths such as those used for maritime HF SSB radio telephone (2.1 kHz) will not support this waveform.

For our application we have chosen Eudora™ for Windows as the user "client" terminal e-mail software This is a user-friendly e-mail program which is in widespread use in

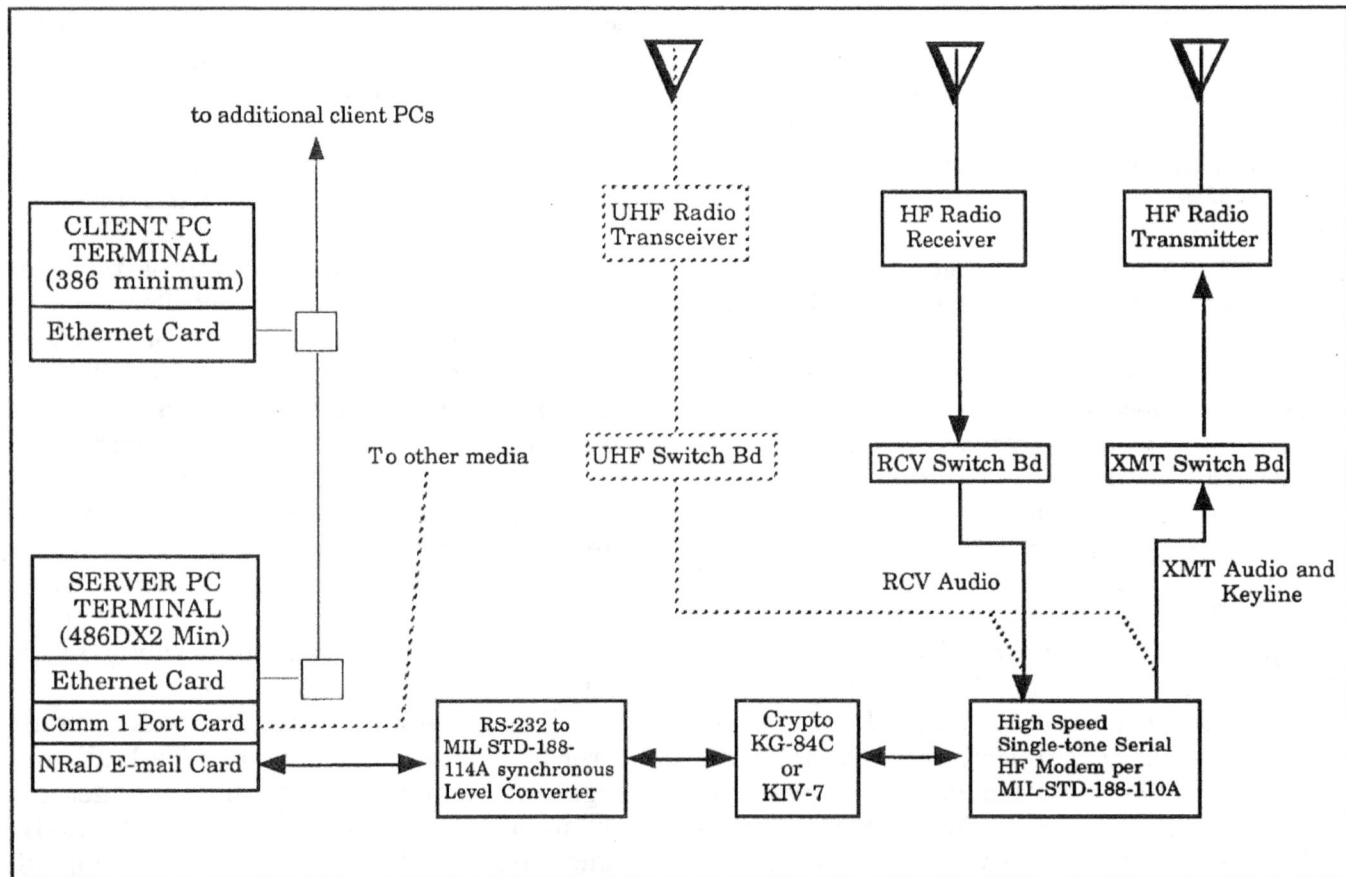

Figure 1. Battle Force e-mail basic system shipboard block diagram.

business and government offices and has an easy-to-use binary file attachment feature. The connection from Eudora to JNOS is via Ethernet, with the simplified mail transfer protocol (SMTP) and the post office protocol (POP3) used to send and receive mail respectively. JNOS netware is based on the open system interconnection (OSI) model, utilizing the Internet standard TCP/IP protocols.

Following closely on the LINCOLN BG success, we installed a similar system on board the USS *Dolphin*, a small research submarine. This system was different because it was to provide long-range ship-to-shore connectivity, with direct access to the public Internet supporting academic research. To aid in establishment of long-range HF channels, we installed a radio transceiver with an Automatic Link Establishment (ALE) feature along with an e-mail server and client terminal. An identical system was set up at NRaD, with an e-mail connection into a mailhome on NRaD's LAN that provided the Internet access. With the simple-to-operate HF ALE radio system, the Dolphin easily "checked her mail" two or three times a day at ranges out to 1,250 miles, without any requirement for an operator at the NRaD end, and only minimal operator effort aboard *Dolphin*. As a result of this success, the *Dolphin's* parent command, Submarine Development Group One, has chosen to obtain similar HF equipment suites for several of their surface support craft. These will be used where possible to reduce INMARSAT operating costs for e-mail service during research deployments, which has run as high as $40,000 for a month at sea.

Building on knowledge gained from the HF application, NRaD has improved the capabilities of the software for use on other media, and has reduced the size of the package to a single PC. This capability was demonstrated during JWID-95 using an ultra-portable system intended for field use by Disaster Area Surveillance Teams for humanitarian relief. This set can operate over HF, VHF and cellular links, and can support standard telephone modem e-mail protocols (such as TELNET, PPP, UDP and SLIP) OVER cellular or other full-duplex links, and common half-duplex e-mail protocols (SMTP and POP) over Ethernet connections and the radio half-duplex links. A more recent

version for military applications was assembled in late 1995. This system uses a single PC, KIV-7 crypto and Rockwell MDM-3001 to form an extremely compact system.

HF e-mail or "Battle Force e-mail" provides the fleet with an inexpensive, simple method for exchanging data among the ships in a carrier battle group, an amphibious ready group, or between groups of a Battle Force — a true at-sea Metropolitan Area Network (MAN) capability. CNO N61 has chosen the USS BELLEAU WOOD Amphibious Ready Group (ARG) to receive an HF e-mail installation in the near future, with the added requirement that a portable system be assembled for the embarked Marine Expeditionary Unit (MEU) for ashore use with their HF radios for an over-the-beach link back to the ARG for e-mail data exchanges. The system on the LINCOLN BG was cross-decked to the NIMITZ BG in Nov '95 for use during her deployment, and installation is in progress for the ENTERPRISE BG and SAIPAN ARG. We anticipate outfitting of other BG's and ARG's prior to deployment.

Recent RDT&E effort has resulted in a relatively simple method or providing "hub-spoke" e-mail communications, whereby the smaller units in the BG or ARG can send e-mail to the "big deck" ship for automatic relay back over their SHF SATCOM links to stateside parent commands via SIPRNET. Future RDT&E effort will look at establishing a bulletin board-like "electronic mailhome" on the big deck for intermittent participants (such as submarines attached to the BG, or MEU units on the move). A more sophisticated routing system is currently being developed in which e-mail and other data will automatically be sent via the best available path, whether it be a high data rate UHF line-of-sight system, HF if the recipient is out of UHF range over the horizon, or satellite links if available. We are also planning an experimental local system that would permit ships operating in Southern California waters to have access to the public Internet via a fixed-frequency station operating at NRaD. This would provide ship-shore e-mail service for ship's business with other commands, and possibly "sailor e-mail" to families with on-line service access. Radio access to this system may be expanded to include UHF line-of-sight and HF ALE as well.

Battle Force e-mail is the opening salvo in providing more information age tools to the fleet. The ability to exchange a variety of information over HF e-mail was quickly recognized by Fleet operators and Marines embarked with a MEU as a valuable asset for planning operations while enroute to an operational objective.

Battle Force e-mail, especially in its ultra-portable form, potentially can be used with forward deployed ground forces, onboard U.S. Air Force transport and tanker aircraft, Coast Guard vessels, and other civil and military users. In a test sponsored by HQ AFC4A, an air-ground link using HF ALE radios provided near-continuous public Internet e-mail access for a USAF transport aircraft while enroute cross-country, using the HF ALE gateway station at NRaD. Battle Force e-mail will likely host the Defense Messaging system (DMS) e-mail applications as a tactical terminal in the near future.

18: TEST REPORT: NEAR VERTICAL INCIDENCE SKYWAVE COMMUNICATIONS DEMONSTRATION

June 24, 1994

by Edward J. Farmer, P.E.

Overview

High Frequency radio communications using Near Vertical Incidence Skywave (NVIS) techniques were demonstrated from a typical narrow mountain canyon — this one near the town of Washington which is approximately 12 miles east of Nevada City. Communications were demonstrated with stations located at OES Headquarters in Sacramento, and with California State Military Reserve (CSMR) stations in East Sacramento, Vallejo, and Simi Valley.

NVIS techniques that were demonstrated included:

Frequency selection in concert with propagation conditions. Antenna selection and installation. Antenna orientation (horizontal vs. vertical).

Digital communications using the PACTOR mode were demonstrated between Washington Canyon and OES Headquarters, Washington Canyon and Simi Valley, and OES Headquarters and Simi Valley.

Reduced power operation (as little as 5 watts) was successfully demonstrated.

All results were as anticipated and were consistent with NVIS principles.

Operational Factors

Frequency selection

In NVIS applications, frequency selection is determined by solar activity (measured by the solar flux index or SFI) and by geomagnetic activity (categorized by the geomagnetic activity indices, mainly the Boulder K index). Solar flux depends on the 11-year sunspot cycle and can range from about 65 to over 200 with higher numbers being better for radio propagation. The Boulder K index ranges from 0 to 9 with conditions being best at 0. For a complete explanation of frequency selection criteria, see Reference 1.

This exercise was conducted during summer conditions (June 24, 1994). During the exercise, conditions were stable with the solar flux index at 73 and the K-index at 0.

SFI and K-index data are computer analyzed to obtain Maximum Usable Frequency (MUF) data for each hour of the day. For this test, these data were analyzed using MINIPROP PLUS, Version 2.0. Data for Washington Canyon to OES Headquarters are shown in Table 1. Note that this table is in UTC. The test period ranged from 2200 UTC (1500 hrs local) to 0400 (2100 hrs local). Maximum usable frequency data for this path are shown in Figure 1. This Figure also shows an estimate of the Frequency of Optimum Traffic (FOT) as well as the frequencies used during this test.

Frequency selection depends on the time of day. In general, higher frequencies work better during daytime because there is less noise and absorption. At night, lower frequencies work better because both the MUF and absorption decrease. This was demonstrated during the exercise. In the afternoon, a test disclosed that propagation was not possible on frequencies as high as 3343.5 kHz. In early evening propagation at that frequency improved markedly and it was even possible to communicate in the medium frequency band at 2490.5 kHz.

Antenna selection

Four antenna situations were used at the field site during this test.

The primary antenna was an approximately resonant half-wave tape dipole. The center was approximately 30 feet above ground and the ends were approximately 20 feet above

Table 1. Data for Washington Canyon to OES Headquarters

MINIPROP PLUS SHORT-PATH PREDICTIONS   6-23-1994   Path Length: 149 km

SSN: 15.7   Flux: 73.0   Radiation Angle: 75 deg for minimum number of F hops: 1

TERMINAL A: 38.51 N 121.49 W   Sacramento   Bearing to B: 23.9 deg

TERMINAL B: 39.74 N 120.78 W   Washington Canyon   Bearing to A: 204.3 deg

Terminal A Sunrise/Set: 1247/0329 UTC   Terminal B Sunrise/Set: 1241/0330 UTC

K: 0

----------------------Signal levels in dB above 0.5 uV-----------------

UTC	MUF	3.0 MHz	3.3 MHz	4.5 MHz	5.1 MHz	6.0 MHz	7.0 MHz	8.0 MHz
0000	5.7	64 A	64 B	60 A	60 A	60 C	60 D	
0030	5.7	65 A	61 A	61 A	61 A	61 C	60 D	
0100	5.8	67 B	62 A	62 A	62 A	61 C	61 D	
0130	5.8	63 A	63 A	63 A	62 A	62 C	61 D	
0200	5.8	65 A	64 A	64 A	63 A	62 C	62 D	
0230	5.8	66 A	66 A	64 A	64 A	63 C	62 D	61 D
0300	5.8	67 A	67 A	65 A	65 A	64 C	63 D	62 D
0330	5.8	69 A	68 A	66 A	65 A	64 C	63 D	62 D
0400	5.8	70 A	69 A	67 A	66 A	65 C	63 D	62 D
0430	5.7	71 A	70 A	68 A	67 A	65 C	64 D	63 D
0500	5.6	71 A	70 A	68 A	67 A	65 C	64 D	
0530	5.4	71 A	70 A	68 A	67 B	65 C	64 D	
0600	5.3	71 A	70 A	68 A	67 B	65 D	64 D	
0630	5.3	71 A	70 A	68 A	67 B	65 D		
0700	5.1	71 A	70 A	68 A	67 B	65 D		
0730	4.9	71 A	70 A	68 B	67 C	65 D		
0800	4.7	71 A	70 A	68 B	67 D	65 D		
0830	4.6	71 A	70 A	68 B	67 D			
0900	4.4	71 A	70 A	68 C	67 D			
0930	4.3	71 A	70 A	68 C	67 D			
1000	4.2	71 A	70 A	68 D	67 D			
1030	4.1	71 A	70 A	68 D	67 D			
1100	4.1	71 A	70 A	68 D	67 D			
1130	4.1	71 A	70 A	68 D	67 D			
1200	4.3	70 A	70 A	67 C	66 D			
1230	4.5	69 A	69 A	66 B	66 D			
1300	4.8	68 A	67 A	66 B	65 C	64 D		
1330	5.0	67 A	66 A	65 A	64 C	63 D		
1400	5.3	65 A	65 A	64 A	63 B	63 D		
1430	5.5	64 A	64 A	63 A	63 A	62 D		
1500	5.6	63 A	63 A	62 A	62 A	61 C	61 D	
1530	5.7	66 A	61 A	61 A	61 A	61 C	60 D	
1600	5.8	65 A	60 A	61 A	61 A	60 C	60 D	
1630	5.8	63 A	64 B	60 A	60 A	60 C	60 D	
1700	5.9	62 A	63 A	59 A	59 A	60 C	59 D	
1730	6.0	61 A	62 A	59 A	59 A	59 C	59 D	
1800	6.0	61 A	61 A	58 A	59 A	59 B	59 D	
1830	6.1	60 A	61 A	58 A	58 A	59 B	59 D	
1900	6.1	59 A	60 A	58 A	58 A	58 B	58 D	
1930	6.2	59 A	60 A	57 A	58 A	58 B	58 D	
2000	5.9	59 A	60 A	57 A	58 A	58 C	58 D	
2030	5.7	59 A	60 A	57 A	58 A	58 C	58 D	
2100	5.6	59 A	60 A	58 A	58 A	58 C	58 D	
2130	5.6	60 A	60 A	58 A	58 A	58 C	59 D	
2200	5.6	60 A	61 A	58 A	58 A	59 C	59 D	
2230	5.6	61 A	62 A	59 A	59 A	59 C	59 D	
2300	5.7	62 A	62 A	59 A	59 A	59 C	59 D	
2330	5.7	63 A	63 A	60 A	60 A	60 C	60 D	

Signal levels suppressed if below -10 dB or if predicted availability is zero.

Availabilities A: 75 — 100%   B: 50 — 75%   C: 25 — 50%   D: 1 — 25%

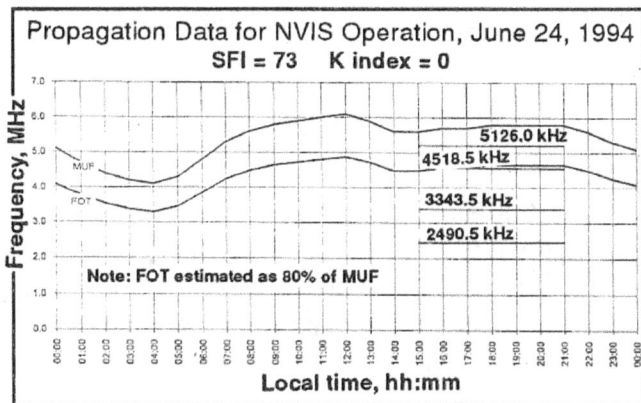

Figure 1. Propagation conditions

ground. It was fed with 75 feet of RG-58 co-axial cable. All operation on this antenna used an ICOM IC-735 100-watt transceiver with an MFJ mobile antenna tuner.

The second, third, and fourth antennas were used in conjunction with an AN/PRC-47 transceiver in order to demonstrate mobile operation. The radio was located on top of a Jeep Cherokee. A ground reference was provided by draping four counterpoise wires over the vehicle.

The second antenna was a 15-foot whip in an approximately horizontal position. The radio was inclined somewhat, placing the antenna's angle of departure at about 30 degrees. In this configuration the tip of the whip was about 3 feet above the earth.

The third antenna used the same radio, location, and counterpoise, however the antenna was in the vertical position.

The fourth antenna used the same radio, location, and counterpoise, however the antenna was a 45-foot longwire which was run through some bushes and was about 3 to 4 feet above earth.

Procedure

See the "Operation Plan" in Appendix A. This plan was implemented as drafted except Ron Fisher, KE6FIN, replaced Bill Spicer, KN6MU. The field team was augmented by Dick Philo, KB6ZCU. The solar flux index was 73 instead of 72.

The field team departed Sacramento at 1200 hrs (local) as planned and arrived on site in Washington Canyon at 1400 hrs. The AN/PRC-47 was set up and the dipole antenna was

erected. Both radios were ready for operation by 1500 hrs.

OES Headquarters kept a log of events and signal strengths as monitored by and reported to them. This log is in Appendix C.

At 1500 hrs. Washington Canyon (Y2L30), made initial contact with OES Headquarters on 5126.0 kHz using the AN/PRC-47 with the whip antenna in the horizontal position.

OES indicated the field unit's signal was Circuit Merit (CM) 3. See Appendix B for a description of the Circuit Merit system. Contact was attempted with East Sacramento (Y2L7). That station was unreadable at OES and weak but readable at Washington Canyon. The East Sacramento station had antenna limitations which effected its operation during the entire test.

At 1510 hrs. Washington Canyon operation was shifted to the IC-735 with dipole antenna. OES indicated this contact was CM5 when the field unit was running 100 watts and CM3 with the field unit running 5 watts.

At 1514 hrs. operation was tested at 4518.5 kHz. OES reported that both the field unit and East Sacramento were CM1.

At 1516 hrs. operation was shifted to 3343.5 kHz. There were no contacts among the stations.

At 1531 hrs. operation returned to 5126.0 kHz. The AN/PRC-47 was operated with the whip antenna in both the horizontal and vertical positions. With the antenna horizontal OES reported CM1. With the antenna vertical OES reported CM0. A standard PRC-47 longwire antenna (45 feet) was rigged approximately 4 feet above ground. OES reported CM2.

Between 1600 and 1930 hours several PACTOR links were attempted. There were some equipment and operational difficulties. At the completion of this work Washington Canyon and OES had linked and passed data, Washington Canyon and Simi Valley (03C5) had linked and passed data, and Simi Valley and OES had linked and passed data. It was possible to maintain a link at power levels as low as 2 watts at Simi Valley and 5 watts at Washington Canyon.

At 1930 hrs. another set of frequency excursion experiments was initiated. The frequencies used included 5126.0 kHz, 4518.5 kHz, 3343.5 kHz, and 2490.5 kHz. The results were:

All stations could communicate on 5126.0. kHz. In the post-test analysis for this series of frequency changes, East Sacramento and OES felt this was the best operating frequency.

All stations could communicate on 4518.5 kHz. In the post-test analysis Vallejo (Y2L3) felt this was the best operating frequency.

East Sacramento was unable to operate at 3343.5 kHz. Vallejo had difficulty loading the antenna but could be heard. In the post-test analysis Simi Valley felt this was the best operating frequency.

The Washington Canyon unit, OES, and Simi Valley were able to make contact and communicate on 2490.5 kHz. While communication was not difficult, this frequency was not as good as the others. Since the sun had not yet set, this result was expected. This frequency would have become progressively better as the sun went down, thus decreasing absorption.

The operation was secured at 2027 hrs. The Washington Canyon field unit returned to Sacramento.

Discussion of Results

All results were substantially as expected. In general, frequencies near the MUF worked better than lower frequencies. Frequencies near the MUF worked very well. Lower frequencies became serviceable as the sun went down.

In the mobile unit tests, the performance of the horizontal whip was better than the vertical whip. The long wire, even though it was strung through bushes at a height of about 3 feet, performed quite well — significantly better than either of the two whip antenna orientations.

Propagation conditions during the test were representative of summer conditions during the low portion of the 11-year sunspot cycle. At 73, the SFI was low, but not as low as it can be. While a lower SFI would have had a negative impact on the test it would not have been remarkable. The geomagnetic activity index (the Boulder K-index) was very low and this contributed to the excellent re-sults. As the K-index increases, propagation becomes more uncertain. Under worst case conditions (e.g., K = 9) frequency selection would have been limited, especially late at night

This exercise was successful because propagation conditions were properly considered. Operating frequencies and antenna types were selected in accordance with prevailing propagation conditions and NVIS communication practices.

Participants

Field Unit; call sign Y2L30; Washington Canyon

Edward J. Farmer, P.E., AA6ZM; Test coordinator

Richard A. Philo, KB6ZCU

Ron Fisher, KE6FIN (CDF VIP)

OES Headquarters; call sign "OES Headquarters"; 2800 Meadowview Road, Sacramento

Stanly E. Harter, ACS/RACES Coordinator

Les Ballinger

CSMR Unit 03C5; Simi Valley, California

Major Steven P. Hall, CSMR

CSMR Unit Y2L3; Vallejo, California

CWO Roger Leone, CSMR

CSMR Unit Y2L7; East Sacramento, California

1 SGT Ronald Fiskum, CSMR

Observer

Fire Captain Specialist Charlie Jakobs, California Department of Forestry.

References:

[1] Farmer, Edward J.; "NVIS Propagation at Low Solar Flux Indices"; **Army Communicator** magazine; Spring 1994; Ft. Gordon, GA.

[2] Shallon, Sheldon C.; **MINIPROP PLUS** computer program and user manual; 1993; Los Angeles.

Appendix A

Operation Plan
OES / CDF NVIS Demonstration
June 24, 1994

Mission

Demonstrate the feasibility of NVIS techniques for providing communication from narrow mountain canyons.

Participants:

California State Governor's Office of
  Emergency Services (OES)
2800 Meadowview Road
Sacramento, CA 95832

California Division of Forestry (CDF)

California State Military Department
State Military Reserve Stations

Y2L30	The field unit
Y2L3	Vallejo
Y2L7	East Sacramento
03C5	Simi Valley

Propagation Conditions

Based on conditions reported for the 12 hours preceding 0900 hrs (local) 23 June 1994, the following conditions are expected during the test period: SFI = 72; K-Index= 0

Figure 1 illustrates the propagation conditions expected. Table 1 provides path probabilities. Note that the time index for Table 1 is UTC. UTC is PDT plus 7 hours. Table 1 was computed for Sacramento-Stockton but is valid for all paths of approximately that length (72 km) or longer.

Mobilization Procedure

The field unit will mobilize at 1200 hrs (local) in Sacramento.

The field unit will travel to Nevada City where it will coordinate with an OES/CDF unit (Bill Spicer, KN6MU) on VHF using the Amateur Radio repeater on 146.625(-) PL151.4 which is located in Grass Valley. When the units are in proximity they will switch to Simplex operation on 144.120. If that frequency is not available they will switch to 144.140.

The units will rendezvous on Highway 20 at the Washington turn-off (approximately 12 miles east of Nevada City) and will proceed into Washington Canyon.

A suitable site in Washington Canyon will be selected and field-expedient HF antennas will be erected. It is intended that the field unit will be on the air by 1500 hrs (local).

Operating Plan

To establish initial contact the field unit will

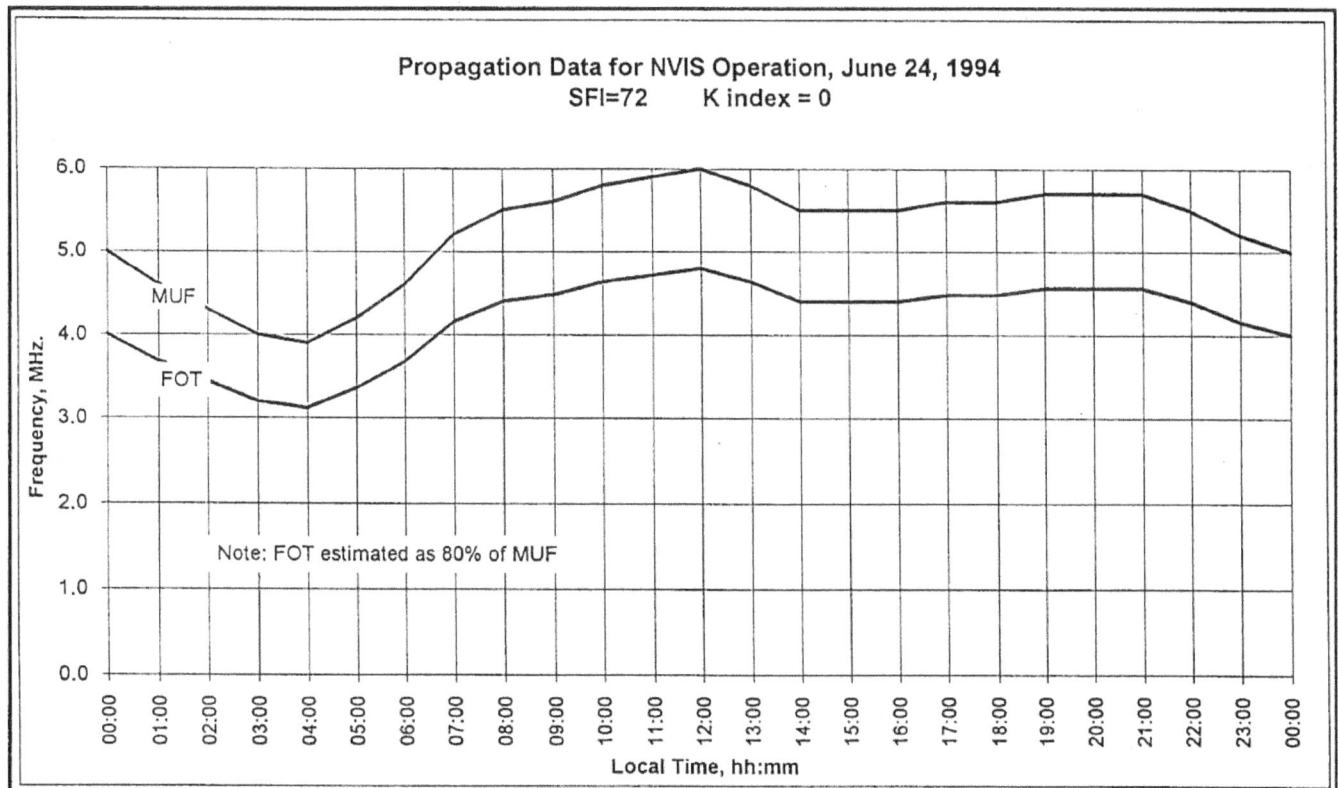

Figure 1. Propagation Conditions

Table 1. Path Probabilities and Signal Levels

MINIPROP PLUS SHORT-PATH PREDICTIONS   6-22-1994   Path Length: 72 km
SSN: 14.3   Flux: 72.0   Radiation Angle: 82 deg for minimum number of F hops: 1
TERMINAL A: 38.51 N   121.49 W   Sacramento   Bearing to B: 162.0 deg
TERMINAL B: 37.89 N   120.24 W   Stockton   Bearing to A: 342.1 deg
Terminal A Sunrise/Set: 1247/0329 UTC   Terminal B Sunrise/Set: 1248/0326 UTC

K: 0
----------------------Signal levels in dB above 0.5 uV-----------------

UTC	MUF	3.0 MHz	3.3 MHz	4.5 MHz	5.1 MHz	6.0 MHz	7.0 MHz	8.0 MHz
0000	5.6	66 A	61 A	61 A	61 B	61 D	60 D	
0030	5.6	41 A	62 A	62 A	61 B	61 C	61 D	
0100	5.6	63 A	63 A	62 A	62 A	62 C	61 D	
0130	5.7	64 A	64 A	63 A	63 A	62 C	62 D	
0200	5.7	65 A	65 A	64 A	63 A	63 C	62 D	
0230	5.7	67 A	66 A	65 A	64 A	63 C	62 D	62 D
0300	5.7	68 A	67 A	66 A	65 A	64 C	63 D	62 D
0330	5.7	69 A	68 A	66 A	66 A	64 C	63 D	62 D
0400	5.7	70 A	70 A	67 A	66 A	65 C	64 D	
0430	5.6	71 A	71 A	68 A	67 A	65 C	64 D	
0500	5.5	71 A	71 A	68 A	67 A	65 C	64 D	
0530	5.3	71 A	71 A	68 A	67 B	65 D	64 D	
0600	5.2	71 A	71 A	68 A	67 B	65 D	64 D	
0630	5.2	71 A	71 A	68 A	67 B	65 D		
0700	5.0	71 A	71 A	68 A	67 C	65 D		
0730	4.7	71 A	71 A	68 B	67 C	65 D		
0800	4.6	71 A	71 A	68 B	67 D			
0830	4.5	71 A	71 A	68 C	67 D			
0900	4.3	71 A	71 A	68 C	67 D			
0930	4.1	71 A	71 A	68 D	67 D			
1000	4.0	71 A	71 A	68 D	67 D			
1030	3.9	71 A	71 A	68 D				
1100	3.9	71 A	71 A	68 D				
1130	4.0	71 A	71 A	68 D				
1200	4.2	71 A	70 A	68 D	67 D			
1230	4.4	70 A	69 A	67 C	66 D			
1300	4.6	68 A	68 A	66 B	65 D			
1330	4.9	67 A	67 A	65 B	65 C	64 D		
1400	5.2	66 A	66 A	64 A	64 B	63 D		
1430	5.4	65 A	64 A	64 A	63 B	62 D		
1500	5.5	63 A	63 A	63 A	62 B	62 D		
1530	5.6	62 A	62 A	62 A	62 A	61 D		
1600	5.6		61 A	61 A	61 A	61 C	60 D	
1630	5.7	66 A	60 A	60 A	61 A	60 C	60 D	
1700	5.8	65 A	59 A	60 A	60 A	60 C	60 D	
1730	5.9	64 A	64 B	59 A	60 A	60 C	59 D	
1800	5.9	63 A	63 B	59 A	59 A	59 C	59 D	
1830	6.0	62 A	63 A	58 A	59 A	59 C	59 D	
1900	6.0	62 A	62 A	58 A	59 A	59 C	59 D	
1930	6.0	62 A	62 A	58 A	58 A	59 B	59 D	
2000	5.8	62 A	62 A	58 A	58 A	59 C	58 D	
2030	5.5	62 A	62 A	58 A	58 A	59 D		
2100	5.5	62 A	62 A	58 A	58 B	59 D		
2130	5.5	62 A	63 A	58 A	59 A	59 D		
2200	5.5	63 A	63 A	59 A	59 A	59 D		
2230	5.5	63 A	64 B	59 A	59 B	59 D	59 D	
2300	5.5	64 A		60 A	60 B	60 D	60 D	
2330	5.6	65 A	60 A	60 A	60 B	60 D	60 D	

Signal levels suppressed if below -10 dB or if predicted availability is zero.
Availabilities A: 75 — 100%   B: 50 — 75%   C: 25 — 50%   D: 1 — 25%

call "Y2L" and "OES Headquarters" on the following schedule:

1500 hrs	5126.0 kHz	USB
1510 hrs	4518.5 kHz	USB
1520 hrs	3343.5 kHz	USB
1530 hrs	5126.0 kHz	USB
1540 hrs	4518.5 kHz	USB
1550 hrs	3343.5 kHz	USB

If contact is not made this same procedure will be repeated each hour until 2100 hrs.

If connection is made Y2L7, or the base station with the best signal, will assume net control duties. The net will select the optimum operating frequency and establish normal net operations.

Operation with the field unit using a vehicular-mounted horizontal whip antenna will be tested.

If a base station with PACTOR capability is available a PACTOR link will be attempted. During the PACTOR experiment an effort will be made to determine the lowest transmitter powers at which the link can be reliably maintained.

The field station will secure on or about 2100 hrs (local) and will return to Sacramento.

Directory:

Test Coordinator: Ed Farmer
Amateur Call: AA6ZM
Military Call: Y2L30
Home Telephone: (916) 393-4066
Office Telephone: (916) 443-8842
Office FAX: (916) 443-3759
Cellular Telephone: (916) 956-8847

State OES: Stan Harter, Les Balinger
Military Call: "OES Headquarters"
Telephone: (916) 262-1603
FAX: (916) 262-1677

OES/CDF Field Coordinator: Bill Spicer
Amateur Call: KN6MU
Telephone: (916) 288-0312

Appendix B

High-Frequency SSB Signal Report System
by Circuit Merit

The Circuit Merit system is used by telephone and HF-SSB radio professionals. The old R-S-T system of reporting in Amateur Radio is vague, inaccurate and complicated. The use of circuit merit better quantifies the quality of a voice signal. Be honest in your reports and report only the average of communications exchange.

CM5 — Completely clear with broadcast quality. Each word fully understood. No objectionable interference or noise. Always breaks squelch*. This designation is seldom earned. Conditions must be superb.

CM4 — Clear with a slight amount of static and/or interference. Each word is understood. Always breaks squelch. The most common report for solid voice communications and very good conditions.

CM3 — Static and/or interference is present. Bulk of transmissions are understood without having to be repeated. Breaks squelch. We deem this to be the margin of acceptable voice communications.

CM2 — Static and interference very prevalent. Words are missed. Retransmissions are necessary. Won't break squelch or squelch is intermittent. This not an acceptable communications circuit and a new frequency must be selected.

CM1 — You can tell only that someone is there but the signal is barely evident and words are unintelligible. Will not break squelch. This is not an acceptable communications circuit and a new frequency must be selected.

CM0 — Absolutely no signal is detectable.

*A syllabic-derived SINAD squelch designed for High Frequency Single-Sideband radio receivers.

Appendix C

Ellen Harter on duty   P.57 PAGE # 1

NOTES:
① Kenwood 9305 on rooftop NVIS 1.8-30 N/S
② Comm done w/ MOTOROLA MICOM-X w/ NVIS B&W broadband 18' above ground
③ Field unit Y2L30, SAC Y2L7, VALLEJO Y2L3, SIMI VLY OJC5.

6-24-94 DATE/ TIME	CALL	5126	4518.5	3343.5	NOISE 5-UNITS ①	
1500	Y2 L30	CM3			(NOT ON UNTIL 1547)	At Washington on the Yuba River. 100W on 15' whip horizontal. Farmer, Philo, Jacobs
1502	Y2 L7	CM1				E.SAC. unreadable.
1510	L30	CM5 CM3				DIPOLE. 100W. REDUCES P₀ TO 5 WATTS!
1514	L30 L7		CM2-1 CM2-1			
1516	L7 L30		CM-1 ∅			
1531	L30 L7	4-5 1				WAITING FOR LES AND PACTOR TEST
1547	L30	5 59+20			7	
	L3	3 5.8			7	
1555	L30	1-∅ ∅			7	PRC on whip HORIZ. NOR VERTICAL
1559	L3∅	∅			6.5	
1600	L3∅	59+10 CM5			6.5	TAPE DIPOLE
1610	L3∅	CM2 56½-7				PRC on Long Wire. Won't break squelch.
1610	L3	CM5 59				
1611	L7	CM∅-1				
1612	L3∅	CM5 59+20			6.5	Going to Monitor mode until 1800
1614	L3∅	1				Wants Les Ballinger ASAP. He is enroute
1650	Note				X	Les Ballinger on duty and hooking up for PACTOR. He needs antenna so no more S-meter readings.
1705	L30					We are ready w/ PACTOR
1709	L3∅	✓				PACTOR TEST STARTS. No good
1721	L3∅	✓				Try again.
1722	Note: Setup 5 radio on LPA:				1	N/G. QSY @ 1728. (S-Unit monitor is 9305 on 6-30 MHz L.P.A.)
1729	L3∅		CM5 S4		3.5	PACTOR. CONNECT! @ 1733. Then drops off. QSY to 5126 + wait for 1800.
1743		MON			<1	
1755	NOTE: We connect both transmitters to the best NVIS antenna through an A-B switch. Improved noise level from 59+10 to S5.					

NOTES:
① Kenwood 930S on rooftop NVIS 1.8-30 N/S
② Comm done w/ MOTOROLA MICOM-X w/ NVIS B&W broadband 18' above ground

DATE/TIME	CALL	CM and S-units ①			NOISE S-UNITS ①	NOTES
		5126	4516.5	3343.5		
1800	L3	59 CM4			7-8	Icom 735 Tape dipole — We are looking at NVIS slopes on roof.
"	L30	CM5 9+10				
"	L7	9				
"	OES4Q	9+10				
"	C5	0				
"	L30	CM4 59+10				PRC47 long wire, 59 to 59+10
1805	L30	CM3 57-58				PRC4y Vertical whip horizontal WBC. Fluctuates.
1811	L30	CM3 57-58			7-8	Reconfigured vertical into vertical. (HUH?) Seems to be a slight improvement with time
1813	O3C5	CM5 9+20+30				Best station yet! State Hall
1815	L30	CM5 9+30				
1820	C5- L30	59+10			8	Pactor. C5 is on LSB !! C5 is 5128.0, L30 is 5126.9. Wow Δf! 200 baud. 25 watts. FCC would be all over us with this Δf!
	"					Now 100 baud 5 watts. Now 200 baud. Somebody now on 2 watts.
1824	"					Now appear both to be on frequency.
1826	"					Back to voice. 3 watts
1828	C5	59+10				Power tests 5 watts
"	L30	59+20				Back to full power on
1831						
1841	C5→OES	+20				Pactor. We cannot link on LSB.
1845	"					We move to USB. He looks good but no link
1850	"					We can copy him but no link.
1900	"					We all discover 10 programming problems to be resolved.
1901	"					LINK ESTABLISHED : 5123.60 KHz
1927	C5	+20 } VOICE				L30 scope pattern / tone distorted.
1927	L30	+20 }			8	S.F. 73 A=2 K=0 15-7 Sunspot
	L3	+30				

NOTES:
① Kenwood 930S on rooftop NVIS 1.8-30 N/S
② Comm done w/ MOTOROLA MICOM-X w/ NVIS B&W broadband 18' above ground

KENWOOD 930-S

DATE/TIME	CALL	CM and S-units ①			NOISE 5-UNITS ①	NOTES
		5126	4516.5	3343.5	①	
1934	L7	Very weak			8	
1946	L30		~~~~~	CM3 CM5+20	8	5 watts / 100 "
"	L7		~~~~~	∅-1	8	
"	L3		~~~~~	CM2	8	
2010	L30		[2490.5]	9+20	9.1	
	C5			9+12	9.1	
2020	L30	9.30			9.0	} much better S/N here @ OES.
	C5	9.30-40			9.0	}
2026		✓				wrap up. Excellent exercise.
2027						I secure. S20 has updated Motorola.

AN/PRC-150 HF radio in urban combat

– a better way to command and control the urban fight

by retired LTC David M. Fiedler and LTC Edward Farmer

Communications in the urban environment

Using Army standard-tactical-radio communications systems on urban and complex terrain has never been very easy. Inherent equipment limitations found in military radios (low power levels and inefficient antennas) coupled with system degrading effects inherent in the urban setting such as signal absorption, scattering and diffraction present many challenges for the combat-net radio user operating with the current suite of military frequencies (2-512 MHz).

Civilian police, fire and municipal service agencies have faced these same challenges for many years. The classical answer has been to position retransmission stations (repeaters) at strategic locations on the urban area of operations. By placing repeaters intelligently (usually atop high structures) and by selecting power levels and antennas with good coverage patterns city governments have long been able communicate among base-station, hand-held and vehicular radios pretty well. As far back as the 1930s the radio frequencies employed by civil government were in the same general very-high frequency/ultra-high frequency range used by many of today's military radios. Recently, in order to relieve frequency congestion and bandwidth availability problems, many urban centers have migrated to much higher frequency ranges where scattering, reflection and absorption are worse than they

are in the military VHF/UHF frequency bands. To compensate, multiple remote repeaters connected to transmission hubs are used to improve coverage over wide areas or into hard-to-cover spots. Network repeaters are connected to command stations (trunked) over telephone cable, fiber-optic cable or microwave carriers and typically assure maximum reliability, area coverage and user access to the civil networks.

Modern cell-phone networks now also operate in this same general frequency range. Each "cell" access point (antenna tower) is positioned for direct (radio line-of-sight) connectivity to subscriber-cell phones located in their coverage area. The access points are interlinked with additional infrastructure including switches and tie lines.

These systems work well in the civil-urban environment because the system designers have the luxury of controlling the infrastructure and major-system parameters such as power levels, antenna locations, number of access points and repeaters. If a "dead spot" is discovered it is usually a simple matter to engineer and interconnect additional repeaters or cells to eliminate it. In addition, most civil-radio and cell-phone communications are directed to subscribers in relatively open non-

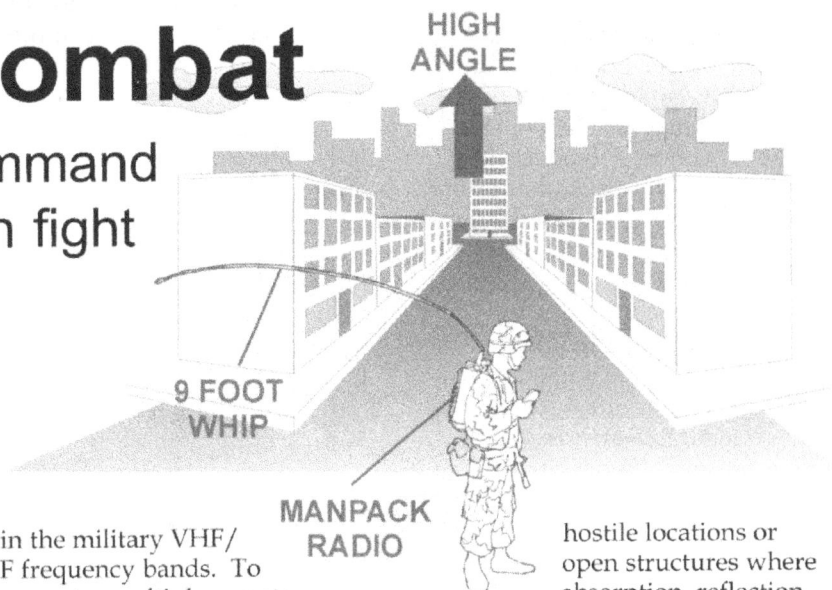

hostile locations or open structures where absorption, reflection and other signal-propagation losses are a factor that can be dealt with. When operation becomes marginal users can simply step outside or move closer to a building window etc. and operations will normally improve as a result of improved line-of-sight signals to the repeater or cell access point.

It's different with the Army...

When the Army is engaged in urban-combat operations the communications situation is considerably different from the situation faced by civil government or cell phone users. Military difference factors include:

1) operation restricted to the frequency range of common military radios (2-512Mhz),

2) limits on the output power of military radio equipment,

3) limited number of available repeater assets if any,

4) limited access to good repeater locations due to enemy action,

5) need to communicate to both outside street locations and inside structures,

6) lack of standard compact antenna systems useful for urban combat,

7) severe restrictions on the

movements of system users,

8) lack of manpower required to cover multiple signal sites can easily exceed available resources. And more.

... but there are ways ...

Fortunately, there are new equipment and techniques available in the force that can, if intelligently applied, overcome many of the communications limitations created by urban combat. One of these is the use of the lower portion of the HF radio spectrum.

Near Vertical Incidence Sky-wave

For many years the Army has known that radio signals in the lower portion of the HF frequency spectrum (2-8Mhz) when radiated at near-vertical angles shower down off the earth's ionosphere (a atmospheric layer of electrically-charged gases at an altitude of approximately 200 miles) in an omni-directional gap-free energy pattern with a radius of hundreds of miles. This transmission technique is called Near Vertical Incident Sky-wave because the signal energy is launched mostly on high (toward the sky) angles between 45 degrees and the zenith and returns to earth after ionospheric reflection. The returning signal comes down from above at high angles in an omni-directional pattern that has no gaps and a radius of hundreds of miles.

While in the past the Army was primarily interested in NVIS for covering theater/corps size areas of operations NVIS is also very useful on the urban battlefield. The advantage of NVIS signals for urban combat is simply that most of the radio energy after ionospheric refraction is not bent, blocked or absorbed by the urban environment in the way that surface wave (low angle) signals from vertical antennas would be. NVIS signal losses are limited to only free space path loss and some absorbs ion at the ionosphere reflection point. Because of this, a Soldier with the Army's new AN/PRC-150 HF man-pack radio (see *Army Communicator* Winter 2001)

and the correct (horizontal) antenna (see *Army Communicator* Fall 2002) can easily receive these high-angle signals if located in open areas between urban structures such as streets, parks, roof tops and other open urban places. The communications path is from the transmitting antenna to the ionosphere and on to the receive antenna. Transmission losses remain fairly constant at around –120 db (a number that can be overcome easily with our equipment) over the entire area covered by the signal. The NVIS signal pattern is truly omni-directional even at very short distances and this makes the transmission mode useful for urban fighting as well as wide area and long distance communications.

HF and structures

Because of their longer wavelengths (lower frequency) HF (2-30Mhz) signals will naturally penetrate urban structures more deeply than signals on higher, shorter wavelength frequencies. How deep the penetration depends on exact frequency, signal power level, antenna efficiency and the makeup of the urban structures in the path.

The name of the game in all radio communications and particularly urban combat radio communications is overcoming path loss. Simply put, the greater the radiated signal and the lower the frequency the more path loss can be overcome. This raises the probability of successful communications in urban areas and inside buildings. Stated mathematically, and greatly simplified:

π is the well-known constant, d is the distance between transmitter

$$Path\ Loss\ (PL) = 20\log\left(\frac{4\pi d}{\lambda}\right) + K_\lambda$$

and receiver, λ is the wavelength at the operational frequency, and $K\lambda$ is a power loss constant determined by characteristics of the obstructions in the signal path at the wavelength of the operational frequency. For

grounded solid-metal buildings without windows etc. K is a very large, meaning that path losses cannot be overcome in order to communicate. For wood and tarpaper structures still found in many urban environments K becomes very small so the first term in the equation predominates. Brick and concrete structures increase K but not to a level where communications fail more often than not. Most structures are inherently (and surprisingly) fairly radio-transparent at HF frequencies. As an example of HF signal penetration it is not uncommon for a small ground penetrating radar transmitter operating in the HF frequency range to penetrate over 100 feet into common kinds of earth while the same power radar on a higher frequency will penetrate much less.

What does this equation mean in practical tactical communications terms? It means, for example, that if we are using a common VHF military radio operating at 30Mhz (lowest frequency for single-channeled ground-to-air radio systems etc.) and replace it with an HF radio like the AN/PRC-150 operating, at say, 5Mhz the path loss drops by 20 decibels (db) because of the way that longer wavelength (lower frequency) signals propagate. In this case lowering the frequency is the equivalent to increasing the power of the transmitter by a factor of almost seven.

Another important consideration for urban combat is raw power. Obviously, the more power you have the more path loss you can overcome and the deeper your signals will penetrate into structures. Common tactical VHF man-pack radios like SINCGARS have a maximum output power of four watts. The AN/PRC-150 HF radio has a maximum output power of 20 watts. That is 7db* more signal power to overcome losses caused by the path, path obstructions, inefficient antennas and other signal consuming factors. Yes the extra power will help you but power relationships are tricky, look at the table below:

4 watts	= 36 dbm*
20 watts	= 43 dbm*
50 watts	= 47 dbm*
150 watts	= 52 dbm*
400 watts	= 56 dbm*

dbm* = decibels above a miliwatt. The db* is a logarithmic unit used to describe a ratio. The ratio may be power, or voltage or intensity or several other factors but in this case it is power (watts). If you do the math you will see that you can measure the difference of two power levels by taking 10 log of their power ratio. If the ratio of power is, for example, two, meaning one radio transmitter is double the power of the other then the difference is 3db. Put another way, for every 3db gained by making a more efficient antenna system or cutting transmission line loss etc., is the equivalent to doubling the transmitter power.

The point here is that often, adjustments to antenna systems or operational frequencies to make an antenna more efficient can produce far more dbs of signal power than simply increasing the raw transmitter power. More power will always help overcome path loss for both NVIS and ground wave systems but many times it is not the best or only answer. If you are already operating at the maximum power that the transmitter can produce then these adjustments do become the only way to compensate for path loss and improve signal penetration in the urban combat environment.

Think "system"

Communications between two radio stations requires that the transmitter power – transmitter antenna gain – receiver antenna gain – receiver performance overcome the path loss between stations. A low-power outstation radio such as a man-pack radio with an inefficient antenna used by forward troops can be "compensated for" to a degree when communicating with a base station that is typically using a higher performance receiver and a more efficient antenna. When the path is reversed, the typically higher-power base-station transmitter and the more efficient antenna again compensates for lower performing combat unit radios in the net. Communications between low-power outstations is much more difficult and may even require retransmission (relay) through a more efficient base station.

In the urban fight, man-pack small unit HF radios, such as the new AN/PRC-150 are extremely portable, but are antenna and power challenged. A high degree of portable NVIS (sky-wave) effect can be obtained when needed by simply physically reorienting standard vertical man-pack or vehicle (whip) antennas to the horizontal plane (see Fig. 2). Direct (surface wave) signals are simpler to generate and use inside structures are also produced from the same antenna by just leaving the antenna vertical.

Communication between two stations by either NVIS (sky-wave) or surface wave transmission only requires that the path loss between them be overcome by the radios and equipment at the ends. Surface wave connectivity while simple to produce is often more difficult to achieve when there are signal robbing surface path obstructions. Surface obstructions can be eliminated under some conditions if the path chosen is sky-wave (NVIS). Do not however rule out the use of surface wave (low angle) signals as a transmission mode in urban combat.

A large station such as a fixed or mobile tactical operations center has the opportunity to erect more efficient antennas and operate more powerful radio equipment thus compensating for some of the system limitations encountered when trying to communicate with typical low powered radios (usually man-packs) carried by combat troops. Highly efficient, large, horizontal-wire antennas are fine for fixed or at the halt, company and higher command-post locations. CPs, have more freedom to select good communications sites even on the urban battlefield. Base-station equipment can make up for much of the system losses caused by having to use low power man-pack radios with inefficient antennas at the fighting locations. The decision to use high-angle or low-angle transmission mode is the call of the combat unit Signal officer. This decision must be made based upon and a knowledge of antennas and radio propagation.

Generally, if the fighting is in the streets and from rooftop to rooftop, C2 elements can standoff from the battle area and control the fight using high-angle (NVIS) communications. If the fighting is inside structures and masked from high-angle signals the C2 element may need to get in close and pump

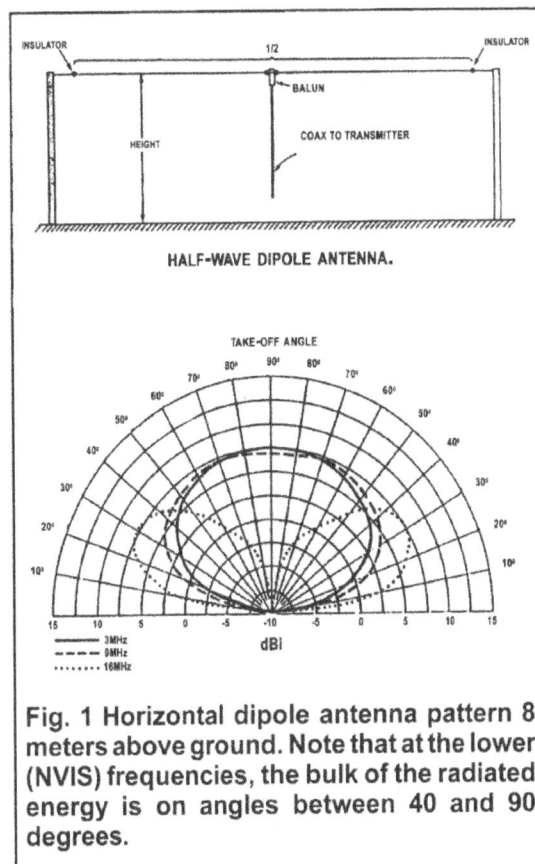

HALF-WAVE DIPOLE ANTENNA.

Fig. 1 Horizontal dipole antenna pattern 8 meters above ground. Note that at the lower (NVIS) frequencies, the bulk of the radiated energy is on angles between 40 and 90 degrees.

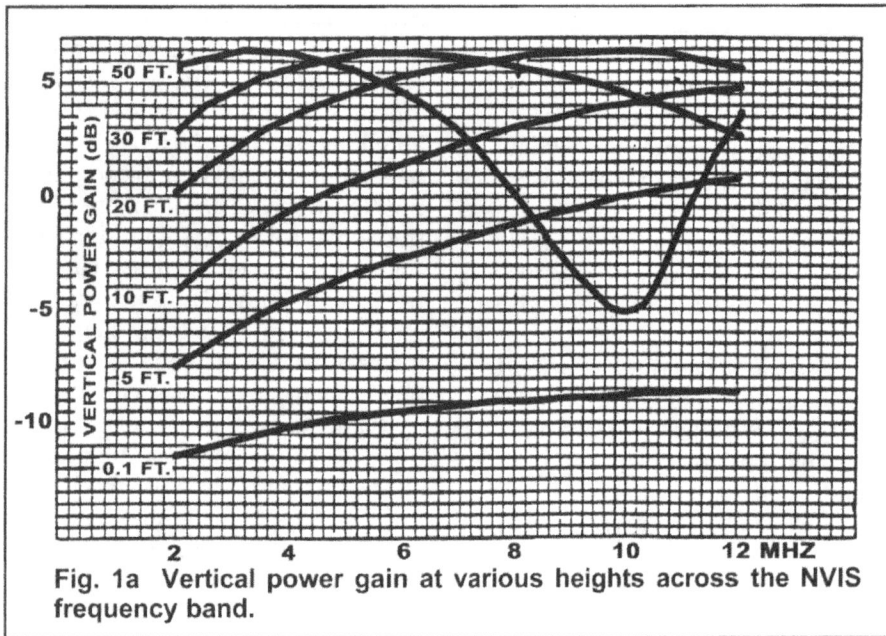

Fig. 1a Vertical power gain at various heights across the NVIS frequency band.

signal energy directly at structures being attacked using vertical (low angle) whip antennas.

A C2 HF base station

High-angle NVIS signals can be easily generated from simple horizontal wire dipole antennas located close to the earth (see Fig.1). The best performance at NVIS frequencies occurs when the antenna is about ¼ wavelength (about 30 feet at 8 mHz) above real ground. The desired gap-free omni-directional antenna pattern shape remains constant, but with markedly reduced signal strength, even when the antenna is lowered to ground level (see Fig. 1a).

A good base-station antenna is critical because it helps the path loss in both directions however, when the tactical situation is such that it is not possible to erect an antenna at the ideal height a lower height will not shut the circuit down. This is true of length also. Perhaps the ideal antenna for a tactical CP base station is the inverted "L" (see Fig. 3a and 3b).

This antenna is efficient if it has the correct dimensions and produces both high angle horizontal polarization for NVIS communications and vertical polarization for compatibility with man-pack and vehicular vertical antennas using low angle (ground-wave) signals at the same time (see *Army Communicator* Fall 02 for discus-

sion on polarization). It is important to note that mixing polarization in Line of Sight ground-wave nets (cross polarization) will cause a huge (20db+) amount of signal loss. Inverted "L" antennas avoid this problem simply because they provide efficient signals with both polarizations in case someone doesn't get the word. Comparing Fig. 1 (dipole) and Fig. 3 (inverted "L") shows the magnitude of the signal difference in the vertical (NVIS) direction when compared to a standard horizontal dipole (Fig. 1).

This loss that is small and is the price paid for generating both high and low angle signals from the same antenna. Inverted "Ls" do need some room to be operated at peak efficiency. Ideal lengths for 35 foot vertical elements are shown below:

frequency range (MHz)	horizontal length (feet)
2.5 – 4.0	150
3.5 – 6.0	100
5.0 – 7.0	80

shorter lengths to match tactical situations will also work but antenna efficiency again will be somewhat reduced.

Portable antennas

The AN/PRC-150 is normally equipped with the OE-505 10-foot vertical monopole whip antenna. Even at ten feet, this is a very "electrically short and inefficient" antenna (an ideal quarter-wave whip at 5 MHz would be 47-feet long). It is normally operated using only the radio loosely coupled to surrounding earth as its counterpoise (radio frequency ground system needed to complete the antenna circuit). This is a very inefficient arrangement compared to what we easily achieve at base stations through the use of balanced antennas (dipoles) or ground radial systems for vertical antennas.

When the fight enters buildings even the ten-foot whip becomes impossible to use. With the full realization that a still shorter antenna will have even lower efficiency than the OE-505 we are left with the requirement to find one. Fortunately, the AN/PRC-150 includes an excellent antenna tuner capable of electrically matching the radio impedance to extremely short antennas, so choices are available.

Physically shortening an OE-505 is an obvious approach, but there's an even better answer that does not destroy the OE-505. The AS-3683 3 foot metal tape antenna that comes with the AN/PRC-119 SINCGARS radio (the most common radio in the Army) will fill this bill perfectly (see Fig. 7). In addition to having a less than 3-foot long radiating element that is short enough to take into a building and stay vertical (the predominant orientation for troops moving inside buildings), the antenna base is a flexible "goose neck" that can be easily bent horizontal for man-pack NVIS operation when the situation permits.

There are some other things we can do to improve the performance both of these admittedly short and inefficient antennas. Operators need to remember that man-pack anten-

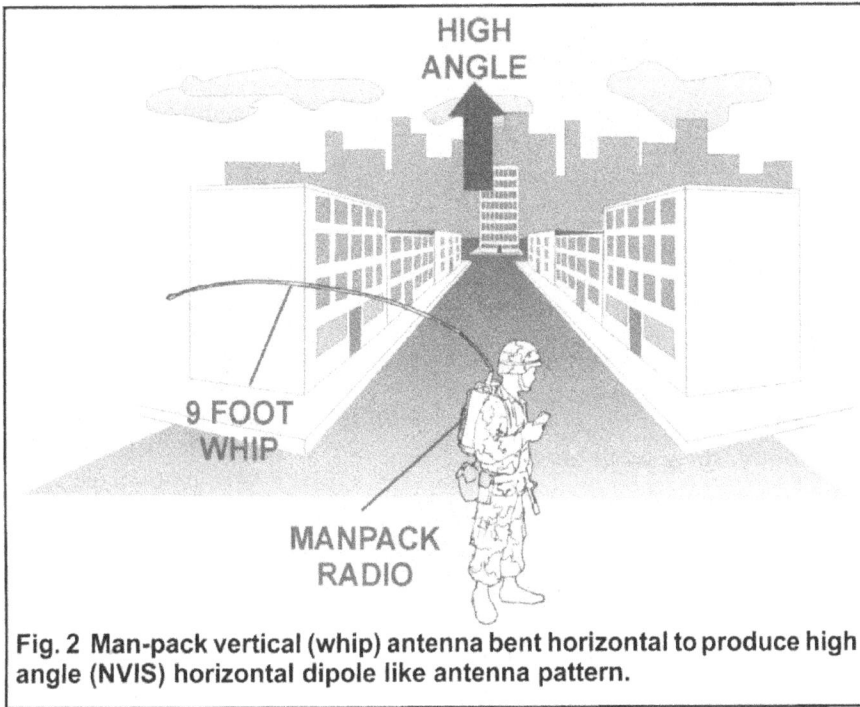

Fig. 2 Man-pack vertical (whip) antenna bent horizontal to produce high angle (NVIS) horizontal dipole like antenna pattern.

nas really consist of the whip (radiating monopole) and what ever is under the whip. All antennas have two sides, and when used in the standard way, the man-pack antennas other side (called the ground plane or counterpoise) really consists of the radio chassis, the operator's body, and whatever the soldier is standing near at the time the radio is operating. Improving the counterpoise/groundplane can provide a tremendous improvement in radiated power and received signal level at almost no cost.

A much better counterpoise in the urban situation is simply a "tail" (see Fig. 6) connected to the radio's ground terminal and hung behind the operator. The longer the tail is the better. Making it about the equal length as the AS-3683 (1 meter) works well in terms of both electrical performance and practicality. Any conductor will do, but the more surface area the better, and copper works better than materials with higher resistive characteristics. The best "tail" construction that we have found is a simple section of computer ribbon cable shorted on both ends with one end terminated on the equipment (chassis) ground. This "tail" can dramatically increase the effective radiated power from the

antenna. When possible, removing the radio from the operator's back will also improve the signal strength since the body will no longer serve as a signal robbing capacitive path to ground. While on the ground, a ground rod and at least four wire radials spread out and connected to the radio ground can produce even greater signal power.

Can it ever get better than this?

This looks great but don't rush off just yet to replace the VHF radio in your squad with an HF man-pack radio. Why? The antenna again! See Fig. 5. The path loss equation above only describes what happens once a signal has been radiated – not how the signal gets generated. You must remember, to

radiate at top efficiency a monopole (whip) antenna should be physically ¼ wavelength (λ) long, and it also needs an extensive low impedance counterpoise. At HF frequencies that is physically a very large antenna. All small antennas suffer inefficiencies.

As an example of how efficiency is reduced as the antenna gets shorter and antenna impedance is mismatched to the radio, look at Fig. 5. Fortunately, modern HF equipment such as the AN/PRC-150 are equipped with a very effective antenna matching unit that is quite capable of providing acceptable antenna electrical impedance matching even to very short antennas. Unfortunately, while the coupling process electrically compensates for a physically short antenna it also reduces effective radiated power of the radio as shown.

The AN/PRC-150 has some additional tricks to help make up for this...

In addition to the higher power levels and better physical

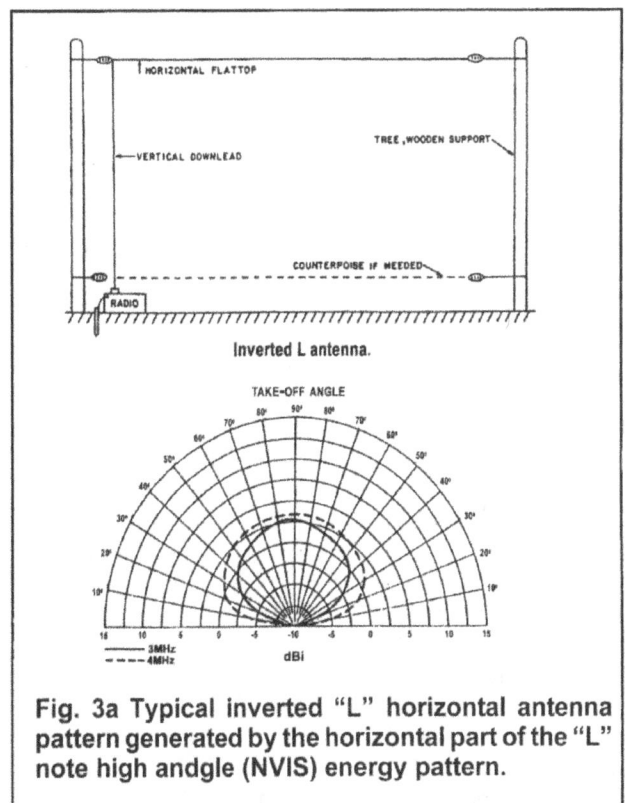

Inverted L antenna.

Fig. 3a Typical inverted "L" horizontal antenna pattern generated by the horizontal part of the "L" note high andgle (NVIS) energy pattern.

signal penetration capabilities of HF radio the AN/PRC-150 has other ways to make back signal lost in the path and the inefficient antenna. The U.S. government (NSA) along with private industry has developed and adopted a new form of digital voice modulation coding called Mixed Excitation Linear Prediction.

MELP implemented in the AN/PRC-150 can operate at both 600 and 2400 bps data rates. MELP has demonstrated an ability to provide a significant increase in secure voice availability over degraded channels particularly at the 600bps data rate when compared to other digital and analog forms of voice modulation. The MELP speech mode uses an integrated noise pre-processor that reduces the effect of background noise and compensates for poor response at the lower speech frequencies. By using digital voice techniques such as band-pass filtering, pulse-dispersion filters, adaptive-spectral enhancement and adaptive noise pre-processing voice communications performance over channels with low signal-to-noise ratios typical of the urban combat environment can now be made useable and reliable.

The MELP capability just like lowering the frequency, using higher power, and improving antenna efficiency translates into dbs of "processing gain" and a better capability to communicate over urban terrain. In effect MELP is compensating for path loss and antenna inefficiency.

The signal-to-noise channel characteristics needed to support various modulation modes are shown in Fig. 8. Note that MELP 600(bps) digital voice performs almost as well as a CW (manual Morse Code) expert operator. Quite an achievement since until recently

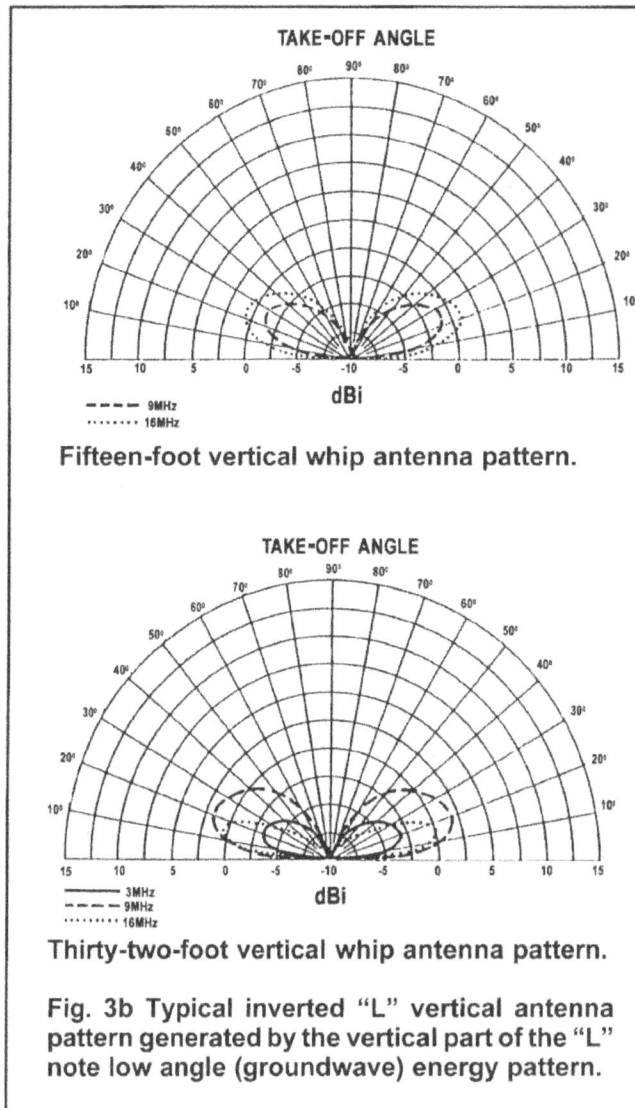

Fifteen-foot vertical whip antenna pattern.

---- 9MHz
.... 16MHz

Thirty-two-foot vertical whip antenna pattern.

Fig. 3b Typical inverted "L" vertical antenna pattern generated by the vertical part of the "L" note low angle (groundwave) energy pattern.

——— 3MHz
---- 9MHz
.... 16MHz

all services tried without success to keep a pool of trained CW operators available because CW Morse Code could get through under conditions that would support no other means of communication. A good look at Fig. 8 also shows this. Analog voice communications is achieved at a S/N ratio of about 12-to-1. Good MELP 600 digital voice communications is achieved at about a ratio of 3-to-1. The ratio of the two modes means a 4-to-1 improvement in communications by going to MELP 600. From the signal power prospective, this is an increase of 6 db (equal to four times the transmitter power) due to gain from digital signal processing. Viewed another way signal gains of this magnitude effectively make a 20 watt radio into the equivalent of an

80 watt radio at the push of a software button but without causing increased stress on radio components that would normally require higher (more expensive) power ratings, and decreased operational life of the radio batteries.

Also shown in Fig. 8 is a digital voice mode identified as Last Ditch Voice. This mode as the name implies is designed to work when nothing else even a manual Morse CW expert will. LDV takes advantage of digital voice processing at a much lower data rate (75bps) in order to slash digital errors caused by marginal conditions. LDV is not a "real time" transmission mode but LDV has both a broadcast and an automatic-request-for-retransmission capability. Voice data packets are created and sent in the transmitting radio. The radio then sends the packets at a very slow data rate using sophisticated error detection and correction digital coding techniques. Data packets are stored in the receiving radio and checked for errors in transmission caused by poor transmission path characteristics. In ARQ mode an automatic request to retransmit corrupted packets can be returned to the transmitting radio in the event to many packets have too many errors for decoding into useable voice communications. In broadcast mode all packets are stored upon receipt the first time. Radio software then assembles the packets and cues the operator. The soldier at the receiving radio then plays the message like a voicemail. The lower data rate and extensive signal processing can produce impressive performance since LDV can recover signals from below the noise levels (see Fig. 8). This again can be equated a considerable increase (perhaps 3db or double) in transmitter power.

To summarize, S-6s and G-6s should consider the following points that make the Army's new family of HF radio a better way to communicate than other means for urban combat if:

1 – Lower signal loss and better penetration into buildings due to propagation characteristics of lower operating frequency.

2 – Higher raw transmitter power to make up for signal losses in the path and due to inefficient antennas.

3 – Lower signal loss through heavy foliage, rain and snow because of longer wavelength.

4 – Lower transmission line losses.

5 – Eliminates need for hard to place and tactically dangerous repeater stations.

6 – Less effected by complex terrain.

7 – Better performance (effective power gain) due to MELP 600 DSP.

8 – Last Ditch Voice digital mode for recovery of extremely weak signals.

9 – Ability to use both sky-wave and surface-wave paths depending on the tactical situation.

Make no mistake; tactical communications under urban combat/complex terrain conditions is sometimes a very hard thing to do. G6 and S6 officers will need to know how to pick an antenna, mode of transmission, and frequency band that will provide the key to success. Much depends upon the skill of unit Signal officers. Using our new HF equipment can help get the message through. Communications planners at every level need to understand the concepts of propagation, path loss, antennas, antenna couplers and digital signal processing as outlined. When they do the chances of getting critical C2 information to all echelons of an urban combat force via HF-CNR will be much better.

Note: At this time, the number of HF radios in the force is not

overwhelming. There will be situations where there just is no AN/PRC-150 or other HF man-pack radios around to use in the urban fight. In this case we will have to fall back on existing stocks of VHF/UHF radios like SINCGARS or AN/PRC-126, or the new commercial-off-the-shelf CNRs that are now appearing in significant numbers such as the AN/PRC-117F (man-pack/vehicular) and the AN/PRC-148 (handheld). The principals outlined above such as using the lowest frequency at VHF and improving antenna efficiency all still apply. Measures

Fig. 5 Effect of shortening vertical HF antennas from 15 feet to 4 feet for convenient operation when in urban combat (6-9db). Shorter antennas will give even greater antenna losses. In addition, most radios will automatically cut output power as the antenna gets shorter.

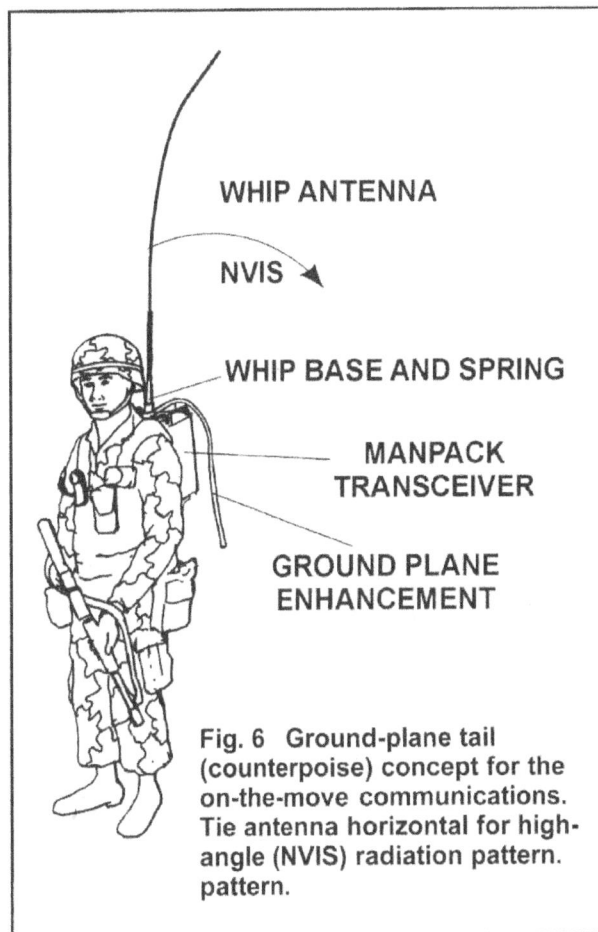

AVERAGE EFFICIENCIES OF 15FT AND 4FT WHIP ANTENNAS IN THE HF BAND

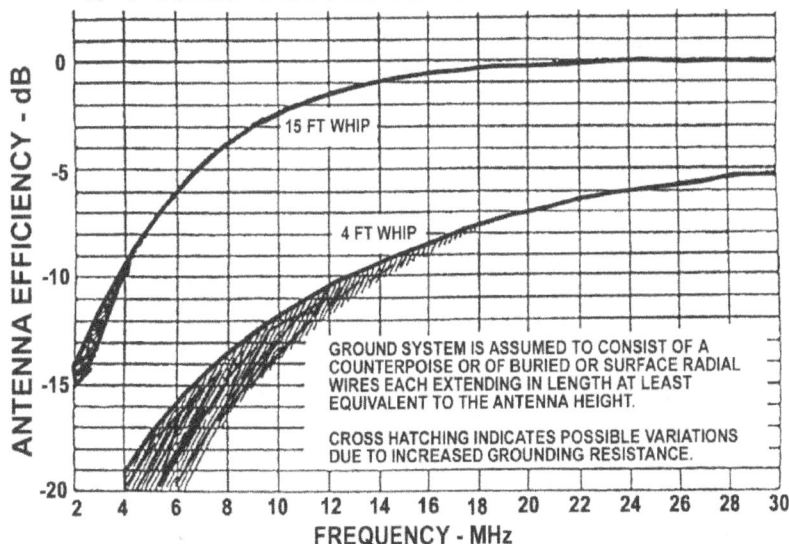

15 FT WHIP

4 FT WHIP

GROUND SYSTEM IS ASSUMED TO CONSIST OF A COUNTERPOISE OR OF BURIED OR SURFACE RADIAL WIRES EACH EXTENDING IN LENGTH AT LEAST EQUIVALENT TO THE ANTENNA HEIGHT.

CROSS HATCHING INDICATES POSSIBLE VARIATIONS DUE TO INCREASED GROUNDING RESISTANCE.

ANTENNA EFFICIENCY - dB

FREQUENCY - MHz

WHIP ANTENNA

NVIS

WHIP BASE AND SPRING

MANPACK TRANSCEIVER

GROUND PLANE ENHANCEMENT

Fig. 6 Ground-plane tail (counterpoise) concept for the on-the-move communications. Tie antenna horizontal for high-angle (NVIS) radiation pattern. pattern.

Fig. 8 Signal-to-Noise Ratios required for different AN/PRC-150 transmission modes. Note LDV recovers signals from below the noise level and MELP 600 operates well in a low S/N ration weal signal environment commonly found in urban combat environments.

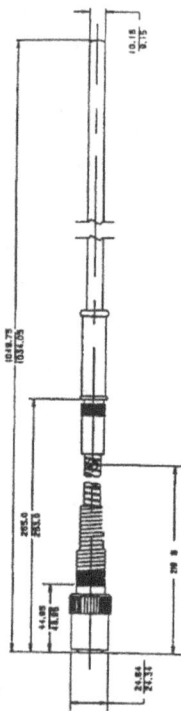

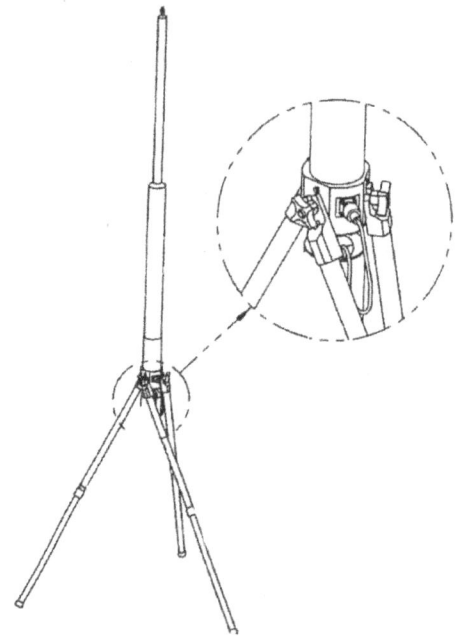

Fig. 7 AS-3683 SINCGARS metal tape Manpack Antenna with flexible gooseneck base that can be made vertical for ground-wave HF communications or horizontal for sky-wave (NVIS) communications. AN/PRC-150 antenna tuner/couplers will impedance match this antenna but efficiency will be poor. Antenna suitable for vertical use inside buildings. Use longer OE-505 if possible. Try higher (shorter wavelength) HF frequencies if possible for better efficiency.

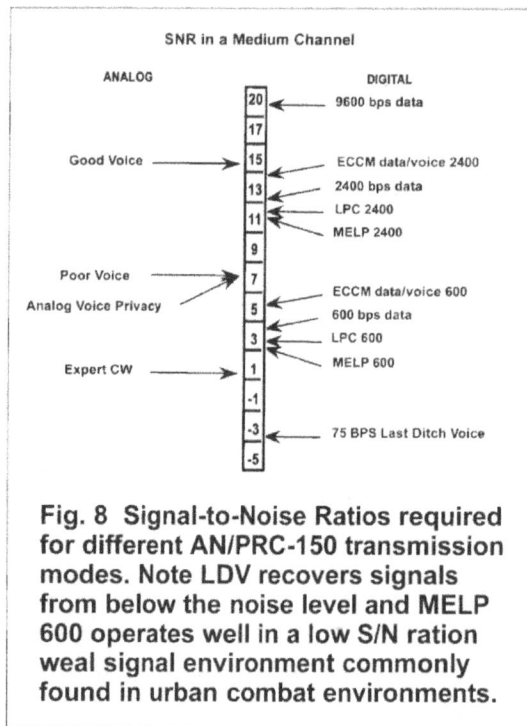

Fig. 9 COM-201 VHF (30-88Mhz) self-supporting ground-plane antenna. COM-201 can easily be brought forward for urban combat since it has a small self-contained package and requires no mast. Good low angle radiation and gain characteristics are a great help under urban combat conditions.

such as providing antenna tails etc. will also help these radios to increase signal levels just like they will an HF radio and for the same reasons. NVIS of course will not apply since the ionosphere cannot reliably reflect high angle signals on frequencies above around 10Mhz or less. If forced to use VHF radios for urban combat there is yet one more thing we can do. Many units are now receiving the COM-201 free standing 30-88Mhz antennas that replace their old OE-254 bi-conical antenna. The COM-201 is an excellent 30-88Mhz vertical ground plane extended range antenna with a low takeoff angle and excellent performance characteristics (see *Army Communicator* Summer 2001). The antenna is designed to be lightweight, easy to move, and to stand on its own

integral tripod/ ground plane. Due to this construction, it is a balanced antenna and therefore more efficient than any man-pack whip etc. The COM-201 can be brought forward and setup on the ground near to where C2 Headquarters are operating or even inside buildings. The combination of high antenna efficiency and low takeoff angle and the use of the lowest possible operational frequency will greatly improve the signal penetration probability for VHF surface wave transmissions. The COM-201 (see Fig. 9) can be connected to any 30-88Mhz radios in the inventory and because of its performance and portability is virtually the only thing in the VHF inventory that can improve standard VHF radio equipment operations in the urban environment. Unit Signal

officers need to be aware of this antenna when only VHF radio is available to support units in urban combat.

Tactical communications using CNR in the urban environment is a hard but not impossible mission for small unit Signal officers. A little basic knowledge about current equipment capabilities and the critical factors of antenna and frequency selection will reduce the difficulty of urban combat communications to a much more manageable task.

To smooth this bump in our professional roads the smart unit Signal officer needs to learn a little, hopefully by reading this article (and other publications) and experiment a lot. Drag out those HF radios and antennas. Even the older ones that don't have all the capabilities of the AN/PRC-150.

Try different antennas, power levels and frequencies etc. until you find the combination of things that work in your situation before you have to do it for real. The same goes for the VHF radios you have. Don't wait to go to the NTC, JRTC or the MOUNT site. The barracks and cantonment areas of major army bases are fine for getting ready to communicate in urban combat. They are just like cities and towns anywhere in the world. National Guard units have it even easier, in many cases all they have to do is get out of the armory and into the neighborhood! The Signal Center also needs to get in gear! Current doctrine, training materials and POIs on how to use CNR in urban-combat just don't have the detail required. Documented requirements for urban combat specific equipment don't exist either as far as we can tell. With the prospects of large-scale urban combat looming larger every day and the reality of Operation Iraqi Freedom with us now, we need to act!

Mr. Fiedler – a retired Signal Corps lieutenant colonel – is an engineer and project director at the project manager for tactical-radio communications systems, Fort Monmouth. Past assignments include service with Army avionics, electronic warfare, combat-surveillance and target-acquisition laboratories, Army Communications Systems Agency, PM for mobile-subscriber equipment, PM-SINCGARS and PM for All-Source Analysis System. He's also served as assistant PM, field-office chief and director of integration for the Joint Tactical Fusion Program, a field-operating agency of the deputy chief of staff for operations. Fiedler has served in Army, Army Reserve and Army National Guard Signal, infantry and armor units and as a DA civilian engineer since 1971. He holds degrees in both physics and engineering and a master's degree in industrial management. He is the author of many articles in the fields of combat communications and electronic warfare.

Mr. Farmer is a Vietnam-era Signal soldier and former lieutenant colonel in California's State Military Reserve, where he ran intrastate emergency communications. He's a graduate of USMC Command and Staff college. He's a professional engineer, has an extra-class Amateur Radio license and is president of EFA Technologies, Inc., in Sacramento, Calif. He has a bachelor's degree in electrical engineering and a masters in physics, both from California State University. He has published three books and more than 40 articles, holds four U.S. Patents and is a frequent guest speaker at communications and antenna-oriented conferences.

ACRONYM QUICKSCAN

ARQ – automatic request for retransmission
CNR – combat net radio
COTS – commercial-off-the-shelf
CP – command post
db – decibels
DSP – digital signal processing
JTRC – Joint Readiness Training Center
LDV – Last Ditch Voice
LOS – line-of-sight
MELP – Mixed Excitation Linear Prediction
MOUT – Military Operations on Urban Terrain
NTC – National Training Center
NVIS – Near Vertical Incident Skywave
PL – path loss
OIF – Operation Iraqi Freedom
RF – radio frequency
SINCGARS – single-channeled ground-to-air radio system
S/N – signal to noise
TOC – tactical operations center
UHF – ultra high frequency
VHF – very high frequency

Mobility favors small antennas:
small-loop high-frequency antennas

by retired LTC Edward J. Farmer, P.E.

In our modern suite of communication options, high-frequency radio has the unique property of requiring no infrastructure. A complete voice and data radio station is easily man-portable and capable, with proper use, of communicating with any other spot on earth.

When the German army was developing the doctrine that became Blitzkrieg it was obvious from the outset that a paradigm shift in communications was essential. Heinz Guderian, the architect of "Blitzkrieg" said, "I want to command over the radio from the front, not talk about it in the rear on a telephone." Since he was originally commissioned as a signal officer and spent much of his career with issues related to staff organization and communication, he had an unusual perspective on the essential roll of communications in maneuver warfare, and how it could be achieved.

A complete HF radio system is

> **Heinz Guderian had an unusual perspective on the essential roll of communications in maneuver warfare, and how it could be achieved.**

easily man-portable, but performance improves with the size of the antenna – and a full-size antenna can be over a hundred feet long. Mobility favors small antennas, and the "holy grail" of HF antenna research is a physically small antenna capable of "full-size" performance. One of the notable efforts along the way, but certainly not the holy grail, is the small loop.

Small-loop antennas have been around for a very long time. While opinions vary as to whether the antennas were loops or top-loaded monopoles, the German army in WWII fielded a number of scout and command vehicles with loop-like antenna structures. Probably the most famous is Erwin Rommel's command vehicle, as seen in Fig. 2.

The idea of a loop antenna comes from the realization that radiation field is the space integral of antenna current over distance. Long antennas with low current produce the same field intensity as small antennas with high

current. The problem becomes designing a radiating structure that promotes the flow of very large radio-frequency currents. The obvious "cut-to-the-chase" answer is, "make a closed loop." If the loop circumference is fairly small its radiation resistance will be small. Because such a structure will be inherently inductive there will be some inductive reactance opposing current flow, but it can be easily eliminated by adding some series capacitance to form a series-resonant circuit. In such a situation, the net reactance is zero and the resistance is the radia-

Fig. 2 General Erwin Rommel's WWII command vehicle showing a loop-like antenna structure.

tion resistance plus the loss resistance of the loop, both of which are very small — perhaps even less than an ohm. This "short circuit" promotes the flow of huge currents and therefore the possibility of large fields from physically small structures.

As the circumference of the structure increases, so does the radiation resistance. Also, the phase of the antenna current in one place is sufficiently different from the phase of the current in another that the radiation pattern becomes a strong function of the frequency of operation, and the expected performance only occurs near the design fre-

Fig. 1 General Heinz Guderian commanding from the front over a radio, circa 1940.

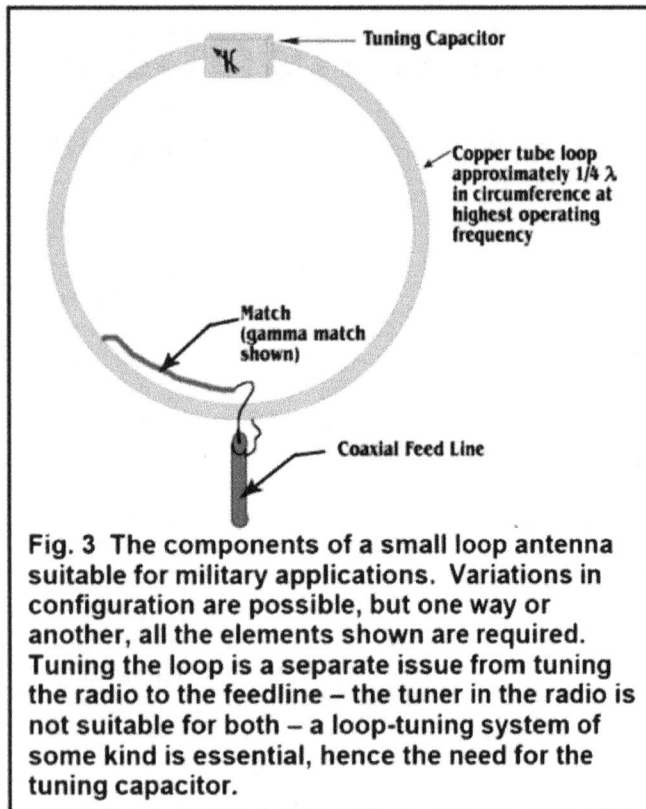

Fig. 3 The components of a small loop antenna suitable for military applications. Variations in configuration are possible, but one way or another, all the elements shown are required. Tuning the loop is a separate issue from tuning the radio to the feedline – the tuner in the radio is not suitable for both – a loop-tuning system of some kind is essential, hence the need for the tuning capacitor.

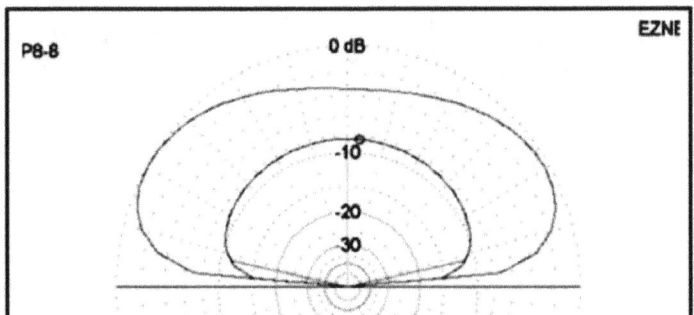

Fig. 4 – a vertical small loop cut for a high frequency of 8 MHz with patterns shown for 2 MHz (inner trace) and 8 MHz (outer trace). Note the NVIS-compatible pattern at both frequencies. In this case there is about 5 dB difference in vertical gain although there is more than 10 dB difference in gain at lower angles.

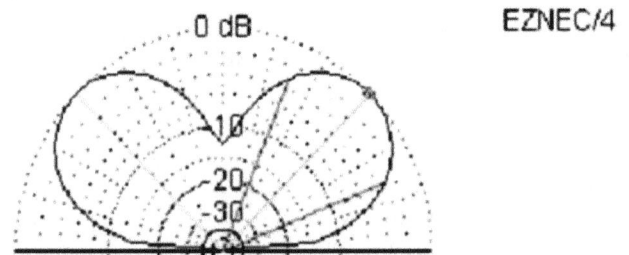

Fig. 5 – The pattern for a horizontal small loop includes an overhead minimum which reduces NVIS effectiveness, but what's missing in the overhead is radiated at lower angels useful for ground wave or long-haul paths.

quency. This causes such a loop to behave more like the linear antennas with which we are more familiar. A classical "full size" loop has a circumference of one wavelength at its intended operating frequency, and isn't especially useful for military purposes.

The "small loop" term is usually reserved for closed-loop antennas in which the current around the loop is more-or-less in-phase, so the loop antenna can be treated as a magnetic dipole. This criteria limits the antenna to a circumference of about ¼-wave-length at the highest frequency at which it is to be used. Also, it becomes harder and harder to match a radio to a small loop as the fre-quency increases – the feedpoint impedance becomes quite large and extremely reactive. Matching a radio to a small loop is one of the very interesting engineering chal-lenges of loop antenna engineering.

The components of a small loop are shown in Fig. 3.

The advantage of a small loop, at least at the high end of its fre-quency range is that it provides gain

and patterns very similar to what one would expect from a full-size (1/2-wavelength) dipole at the same frequency. This is a huge advantage – a physically small, lightweight, easy-to-deploy antenna that provides about the same performance normally ob-tained only after three Soldiers do 15 to 30 minutes work erecting masts and stringing wire.

There are two significant limitations. First, loops are sensitive to objects moving in their vicinity (near field) so re-tuning can be a frequent requirement.

Second, as frequency decreases from the size-defining highest frequency so does efficiency. While a loop will theoretically operate at any lower frequency the efficiency decreases so significantly that practical issues restrict it to about an octave (2:1 frequency range), so the lowest frequency is generally assumed to be about half the highest frequency. While the antenna's pattern remains the same as fre-

quency decreases, the loss in efficiency dramatically reduces the gain. At the lower frequency the loop's gain will be down by about 10 dB from what it was at its highest frequency.

This effectively converts a 100-watt radio at the higher frequency to a 10-watt radio at the lower one, and relegates the performance to some-thing more equivalent to the com-monly used vehicular whip antennas than it does to a full-size dipole. This does not however eliminate the loop from one of its most important military applications, that of a small vehicular on-the-move antenna. It does require that care be taken in trading off antenna size, radiation efficiency, and transmitter power.

Loops can be arranged with the plane of the loop vertical or horizon-tal. Both give satisfactory perfor-mance for modern land HF combat communications. The horizontal configuration produces more lower-

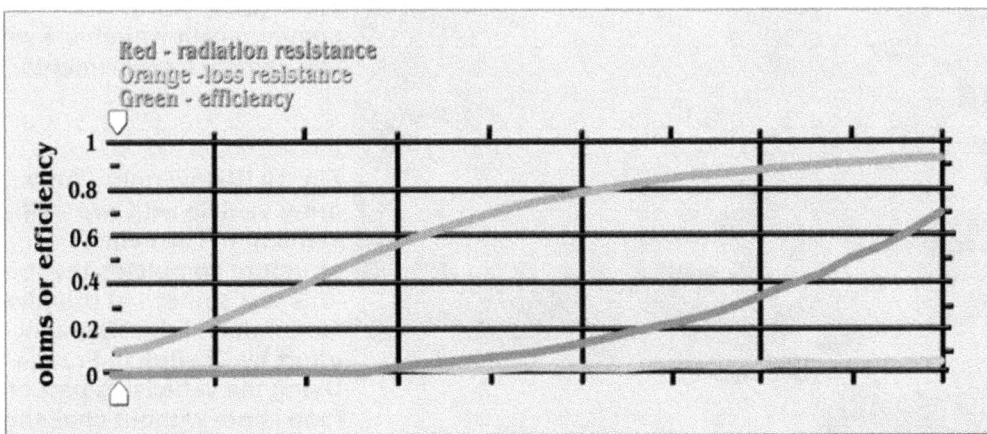

Fig. 6 The graph shows radiation resistance, loss resistance, and efficiency of a 30 MHz small loop over frequency. Note that while efficiency is good at the upper design frequency it becomes less 50 percent just below 15 MHz.

angle radiation useful for long distance (low angle) communication and surface wave (ground) LOS systems, at the expense of the near-vertical radiation required for NVIS (near vertical-incidence skywave) operation. NVIS is the most useful mode for operation in theater/corps-size areas so many of the world's armies (ie. Russia, China and Norway) have opted for this orientation.

The gain and pattern of a vertical small loop is shown in Fig. 4, and a horizontal small loop in Fig. 5.

Understanding efficiency is the key to understanding and effectively using small loops. Assuming the loop-tuning mechanism balances the inductive reactance of the loop itself with the capacitive reactance of the tuning capacitor, then the feedpoint impedance of the loop is the radiation resistance plus the loss resistance. In all radiating structures, radiation resistance increases with length, so we would expect the radiation resistance to be pretty small.

A common relationship for the radiation resistance of a small loop is:

$$Rr = 197 \; [\; Circumference \; /$$

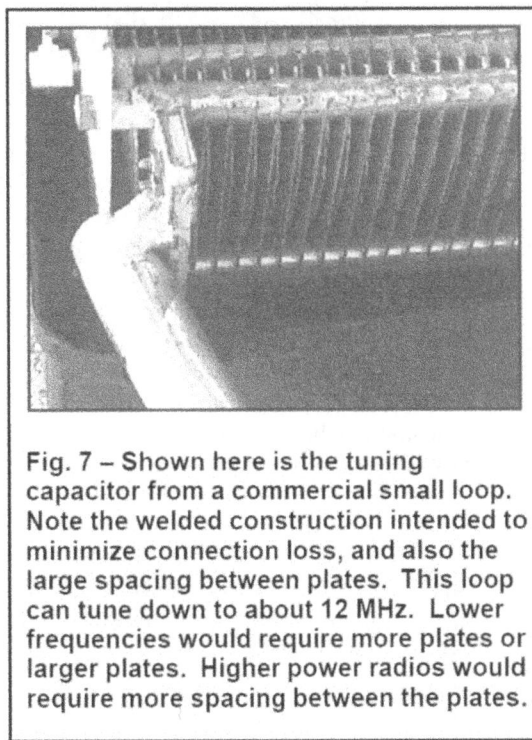

Fig. 7 – Shown here is the tuning capacitor from a commercial small loop. Note the welded construction intended to minimize connection loss, and also the large spacing between plates. This loop can tune down to about 12 MHz. Lower frequencies would require more plates or larger plates. Higher power radios would require more spacing between the plates.

Operating wavelength]4 (eqn 1)

If the loop circumference is ¼ wavelength the radiation resistance is about 0.77 ohms – about a hundredth that of a full-size dipole – but then that's necessary to get the very large loop currents we're after.

If radiation resistance is the "good" resistance (that representing the conversion of applied radio frequency energy into radiation

field) then the "bad" resistance is the "loss resistance." It includes the skin-effect resistance of the loop conductor plus the resistance of all joints and connections. If the connections are kept to a minimum and well-made the main loss is in the tuning capacitor and in the skin-effect resistance of the loop material itself. It is crucial the loop be made of a highly conductive material, and that it be large in cross-section. Assuming the loop is copper, the relationship for skin-effect loss resistance is:

$$Rs = 9.96 \times 10\text{-}4 * \text{"f} * S / d$$

Where R is in ohms
f is the frequency in MHz
S is the circumference in feet
d is the conductor diameter in inches

Note that loss resistance changes as the square root of frequency, while radiation resistance changes as the fourth power of frequency. As frequency decreases from the loop's upper design frequency the radiation resistance decreases as the fourth power of frequency while the loss resistance decreases much more slowly. Efficiency is calculated as:

$$Efficiency = Rr / (Rr + Rloss)$$

Which decreases as the 3.5[th] power of frequency. This is why efficiency falls off so badly near the bottom of the frequency range. This is much easier to visualize on a graph. These data were computed for a loop designed for operation up to 30 MHz and the results are plotted in Fig. 6.

Fig. 8 (Above) A Russian tank with a guard-rail-like small loop. Note the plane of the loop is horizontal.

Fig. 9 (Left) A Russian communications vehicle with two vertical loop elements.

Fig. 10 (Below right) An Israeli army vehicle with two half-loop elements. The vehicle structure completes the loop. This is a somewhat primitive research and development effort by Chelton in France. Using the vehicle as part of the loop is not without challenges. The loss resistance of the vehicle will be much higher than copper or aluminum conductor (the resistance of steel is more than six times that of copper), and the effect of the high radio frequency currents on vehicle components and equipment requires evaluation.

There are two other issues. The first is bandwidth. A radiating structure involving very low resistance and very high reactance is the definition of a high Q circuit, and such circuits have very narrow bandwidth. This means the tuning capacitance will have to be adjusted with even the smallest change in operating frequency.

The second issue is the tuning capacitor itself. It must be adjustable over the required range of values for the specific loop design, and must withstand the substantial voltages (easily several thousands of volts) that appear across it. In the case of "simple" air dielectric variable capacitors (see Fig. 7) this amounts to large spacings between the plates, which, to achieve the required capacitance, involves very big plates.

There are alternatives to air variable capacitors, the two common ones being vacuum variable capacitors (although the large glass enclosure makes them somewhat fragile for military purposes) and discrete component capacitors that are switched in and out of the circuit as needed. The switches have to withstand the very substantial r.f. current flows. One such switch is made by Kilovac Corporation and

amounts to a vacuum relay. Even though fairly small (2-inch diameter) these relays are rated for 25,000 volts and 30 amps making them appropriate for most loop applications.

Whether the tuning capacitor is a rotary (air or vacuum variable) or made of individually switched components operation is much easier if adjusting the loop tuning capacitance for loop resonance is done automatically. This requires some kind of specially designed automatic loop tuner. Such equipment exists and is available in one form or another from loop antenna manufacturers. It is critical to mention, however, that the antenna tuner in a commercial radio probably isn't going to do loop tuning more than once. The high current and especially the high voltage dramatically exceeds the design parameters of these commercial tuners and the odds one would survive loop operation are extremely small. Military tuners and radios will have a somewhat better chance, but the real answer is a purpose-designed loop tuning system.

While there are some challenges with the successful design and application of loops it has been done quite successfully since well

before WWII. The following photographs illustrate some more contemporary applications.

Mr. Farmer is a Vietnam-era Signal soldier and former lieutenant colonel in California's State Military Reserve, where he ran intrastate emergency communications. He's a graduate of USMC Command and Staff college. He's a professional engineer, has an extra-class Amateur Radio license and is president of EFA Technologies, Inc., in Sacramento, Calif. He has a bachelor's degree in electrical engineering and a masters in physics, both from California State University. He has published three books and more than 40 articles, holds four U.S. Patents and is a frequent guest speaker at communications and antenna-oriented conferences.

ACRONYM QUICKSCAN

CNR – combat net radio
GHz – gigahertz
HF – high frequency
JTRS – Joint Tactical Radio System
LOS – line-of-sight
MHz – megahertz
NVIS – Near Vertical Incidence Skywave
Ohm – unit of electrical resistance
R&D – Research and development

HF combat net radio lesson learned again

by retired LTC David M. Fiedler

Recently, at the 2003 Signal Symposium and prior to that in his testimony to the Congress, LTG William "Scott" Wallace, former commanding general of V Corps during the invasion of Iraq, made the following statement about the command, control and communications situation during the Iraq fight. "Despite the introduction of battle-command-on-the-move capabilities that I enjoyed in my assault command post, the vast majority of tactical leaders and CPs (command posts) enjoyed few on-the-move capabilities. Most were tethered to a CP and largely dependant upon line-of-sight communications.

"Case in point. At the corps level the G2 could see individual fighting positions defending a critical bridge because we had a UAV (unmanned-aerial vehicle) leading the lead formations. But we could not get the data down to the unit who was taking the objective because all the CPs were moving. It was a deliberate attack at the corps level, but a movement to contact at the battalion level," Wallace said.

This statement upsets me greatly both as a student of military art, science and history; and as a Signal professional with over 35 years service in all components of the U.S. Army. Wallace's statement when reasonably analyzed can only lead to the conclusion there was a failure in both communications planning and communications execution. The means to provide what Wallace needed (beyond–line-of-sight-on-the-move communications) certainly exist today in our widely deployed family of high-frequency combat net radios and has for many generations. Why then were we not able to improvise, and

Fig. 1 Sd.Kfz-223 wheeled communications/liaison vehicle circa 1935 used for ground to air coordination. note the "frame" (horizontal loop) high frequency antenna that generates NVIS (sky-wave) signals that provide terrain independent radio communications using NVIS signals over corps/theater size areas. Aircraft flying low-level reconnaissance and attack missions passed their information to this facility for further relay via radio, telephone, teletype or messenger to command and control facilities similar to modern U.S. Army Tactical Operations Centers. Communications were self contained and operable both fixed and on-the-move.

adapt our existing resources to overcome Wallace's communications problems?

Wallace and the whole Coalition Force in Iraq were magnificently executing classic offensive "Blitzkrieg" operations. In German, Blitzkrieg means lightning war. In the modern tactical sense it includes attacks where the enemy thinks you cannot attack, rapid advances into the heart of enemy forces and territory, and coordinated massive air and artillery attacks that with today's technology also includes

missiles, attack helicopters and precision guided weapons. The use of such tactics is intended to stun the enemy and shock them to the point that they can no longer react. The German Army in World War II won most of their great victories with this tactic. Field Marshall's Hans von Seekt, Irwin Rommel and Heinz Guderian (a signal branch officer), are all given credit for inventing and perfecting the Blitzkrieg tactic with military scholars giving the lions share of the credit to Guderian the signalman. Guderian was not only

Germany's premier tactician, he eventually became Chief of Staff of the army imagine that happening to a U.S. Signal officer!

Why Guderian from the signal branch? Because, not only was old Heinz a tactical genius who conceived a new combined arms organization to execute the Blitzkrieg concept (the Panzer Division/Corps/Army) but also in his own words circa 1920: "I realized that I would no longer command from the rear with a telephone (*World War I style*) but from the front with a radio". Because of Guderian's signal background and position in the high command, he assured that each tank, aircraft, and unit command post in the Panzer force had long-range, mobile, combat-net radio communications of the right type to support its mission. (See Figs. 1 and 2). The same type radio Wallace needed almost 70 years later.

These were in large part the FuG-10 HF operating in the HF 2-18 MHz frequency range. The Guderian designed HF radio nets provided a level of command and control never before achieved on the battlefield. Long-range (HF) Combat-Net Radio made the Panzer Division and its air support the most destructive and efficient combined arms force in history. The U.S. Army learned much from the Germans of the 1930s and 1940s and thanks to officers such as Fox Connor, Ben Lear, George Marshall, Dwight Eisenhower and the always revered George Patton, the U.S. Army could also combine command and control, logistics, firepower and air support and by 1944 could out Blitzkrieg the inventors of the whole idea. We continue to improve this capability to this day as our victories in Iraq prove.

The basic concept of the German combined arms Panzer force refined by the American Army over the last 70 years and given modern

Fig. 2 Stryker-like Sd.Kfz-232 heavy armored wheeled command-post vehicle circa 1938. Note use of both the NVIS "frame" (horizontal loop) antenna for long-distance wide area on-the-move communications reflected off the ionosphere and a long efficient vertical monopole antenna for shorter distance "ground wave/surface wave" LOS communications. Facility communications were self-contained and capable of on OTM operation. NVIS was the primary mode of OTM omni-directional communications. A short vertical (whip) antenna was also provided for short distance OTM LOS communications.

technology was the force that Wallace entered Iraq with in 2003. In terms of organization and tactics. Rommel and Guderian would have felt quite at home in V Corps with their rapid movements, ability to see the battlefield, and elaborate methods of command, control and communications between ground and air elements. What would have shocked them all but particularly

Guderian with his emphasis on communications, would have been the combat communications failure at the key defended highway bridge that Wallace described, and V Corps' apparent inability to provide timely command, control and intelligence information to its forward elements with the resources it already had. The problem of reaching the battalion Wallace refers to as being unreachable and therefore conducting a movement to contact not a deliberate attack because "all the command posts were moving" is a problem that was solved well before to 1939 in both the German and U.S. armies.

Not only was it solved, it was solved without the use of satellite communications, complex tactical data networks, unmanned aerial vehicles, balloons, retransmission stations etc. A simple single-channel HF radio with the proper antenna, frequency assignment and the knowledge to use it is all that was required both then and now.

By the time the Germans invaded Poland (1939) Guderian had long worked out the techniques of Near Vertical Incidence Sky-wave HF radio communications and how to use FuG-10 HF radios, both monopole and loop antennas, surface wave radio propagation and the reflective properties of the ionosphere (NVIS) to communicate over huge areas when halted, on-the-move, or in the air. (See Fig. 3). Near Vertical Incidence Sky-wave techniques are described in U.S. Army Signal Corps publications as far back as the 1930s and are still currently reflected in our doctrine (FM24-18, FM11-53, FM11-64, FM11-65, TM11-666, MIL-HDBK-413, to name a few). Moreover, ground and airborne HF radio both fixed and on-the-move using NVIS techniques has been a topic of discussion by several authors in the *Army Communicator* more than a dozen times since 1983 alone.

Additionally, Special Operations Forces, Army Aviation and the Army Medical Department have deployed hundreds of new AN/ARC-220 and AN/PRC-138/150 HF radios for exactly this purpose over the last 15 years. C3 elements of the "big" Army also possess large numbers of HF radio's that range from the most modern (AN/PRC-150 family) to somewhat obsolete but still useable (AN/PRC-104 family) but as this instance proves, have not employed them nearly as well. This forces us to raise the question "why were minimum essential doctrinally required HF communications not available to V Corps Headquarters when they needed them in 2003 like they were for Rommel and Guderian in 1939 and for Patton and Eisenhower in the great Louisiana Maneuvers of 1940?"

Further, it also forces us to ask the question why is the U.S. Army with certain notable exceptions (SOF, AMEDD, Army Aviation) the only army in the world and the only department in the U.S. Department of Defense with the mindset to reject a proven viable, inexpensive, means of LOS and BLOS fixed and OTM military communications?

A large part of the answer is that HF communications is the victim of numerous "communications failure myths" created over the years by Signal officers desperately searching for reasons for failure to tell their commanders because they failed to be taught or to learn enough radio technology to effectively use the proven military potential of the HF medium and the equipment they were given. In short, you have to know something about HF to use it. This bad reputation was made even worse in the 1970s when it was coupled with the need to find "bill payers" for other programs such as satellite communications so HF resources were cut for both procurement and training. Additionally, during the same timeframe, Signal Corps leadership was bent on washing its hands of all combat net radio systems by declaring them "user-owned and operated". One can only wonder at the politics

behind this decision. Let's begin to analyze Wallace's problem by debunking some of the worst common myths about HF tactical communications:

Myth 1 – The HF spectrum (2-30Mhz) by international treaty is limited to analog voice single sideband AM modulation in 3Khz channels and therefore can only support digital voice and data at rates no faster than 2400 bps.

-**False;** slow speed digital voice modes are in most modern HF radios for operation over degraded channels but so are MODEMS that can operate at speeds up to 9.6Kbs inside the mandated 3Khz channels. This allows voice, and digital applications such as e-mail, and imagery to be viable HF modes of operation.

Myth 2 – HF radios are not good for short distance tactical communications beyond-line-of-sight and leave gaps in area coverage.

- **False**, while intercontinental communications distances are commonly achieved using some HF techniques, use of properly selected antennas and frequencies will produce antenna patterns good for communications over Corps and below size areas independent of the intervening terrain and without gaps in coverage. ONLY HF RADIO CAN DO THIS WITHOUT THE NEED FOR SATCOM OR UAV SUPPORT! See fig 3.

Myth 3 – HF radio systems are not omni-directional and are therefore not suited for tactical communications.

- **False**, common HF antennas like vertical monopoles (whips), horizontal wire dipoles, and loop antennas all provide omni-directional communications when configured properly for that purpose. Even a horizontal wire dipole when located close to the earth is an omni-directional antenna (see Fig 3). These

antennas can be made directional but only when elevated to a considerable height.

Myth 4 – HF radio systems are more adversely affected by ionospheric storms, solar flares, sudden Ionospheric disturbances and polar blackouts.

- **False to a large degree.** These naturally occurring phenomena to some degree affect all radio systems. The lower portion of the HF band will be affected first. Affects range from almost nothing to complete blackout. Most modern HF systems employed by the Army have a feature called Automatic Link Establishment. This feature scans the radio's assigned frequency band and will automatically establish communications on any authorized workable frequency quickly after a disturbance subsides. Tactical radios operating in other bands and equally affected by these factors don't have these features and may take longer to recover.

Myth 5 – "Sunspots" kill HF radio systems.

- **False to a large degree**. Sunspots are whirling masses of electrically charged gas formed by magnetic fields deep within the Sun. Magnetic fields often more powerful than the magnetic field of the Earth occur at the center of a sunspot. Huge waves of energy produced by the Sun's core erupt through the surface launching a mass of electrified gas and other material. Viewed from Earth this looks like a dark spot on the surface of the Sun where the eruption occurred. The electrified gas has a large magnetic field at its surface that races through space and can disrupt radio communications and electrical systems here on Earth. These disruptions can last a while and affect all radio communications. The lower frequencies such as HF take a while to recover from such disturbances. ALE again will find channels suitable for communications and restore service faster than systems using other tactical radio frequencies. Sunspots don't happen that often but this complex sounding phenomenon has been used to explain signal outages to commanders far beyond what is justified.

Myth 6 – Levels of manmade, atmospheric, cosmic and internal electrical noise are greater in the HF frequency range and cannot be compensated for.

-False, The combination of ALE that selects the best authorized channel based on the best signal-to-noise ratio, higher transmitter power, and the system gain derived from the use of powerful voice and data digital signal processing techniques including Mixed Excitation Linear Predictive coding allow HF communications to proceed in an extremely degraded environment. Some techniques internal to modern army HF radios such as MELP will actually recover signals from near or below the noise level.

Myth 7 – BLOS HF communications OTM don't work.

-False, like everything else in radio system engineering success in OTM/BLOS HF communications depends on the critical selection of antennas and frequency. Vehicle mounted vertical monopole (whip) antennas work and will produce "surface-wave signals". Surface wave signals will propagate out to a certain distance along the earth and then due to their contact with the earth become to weak for use in tactical communications. Depending upon the type of ground or water under the signals, signals at HF frequencies can go relatively short distances to the horizon or in the case of seawater and certain ground conditions tend to bend along the surface of the earth and travel well beyond line of sight. Signals designed to take advantage of ionospheric reflection by using mobile antennas that produce high angle energy (loops and bent over whips) will commonly cover Army Corps/Theater size areas of operations without gaps in coverage. Only signals in the HF frequency band can

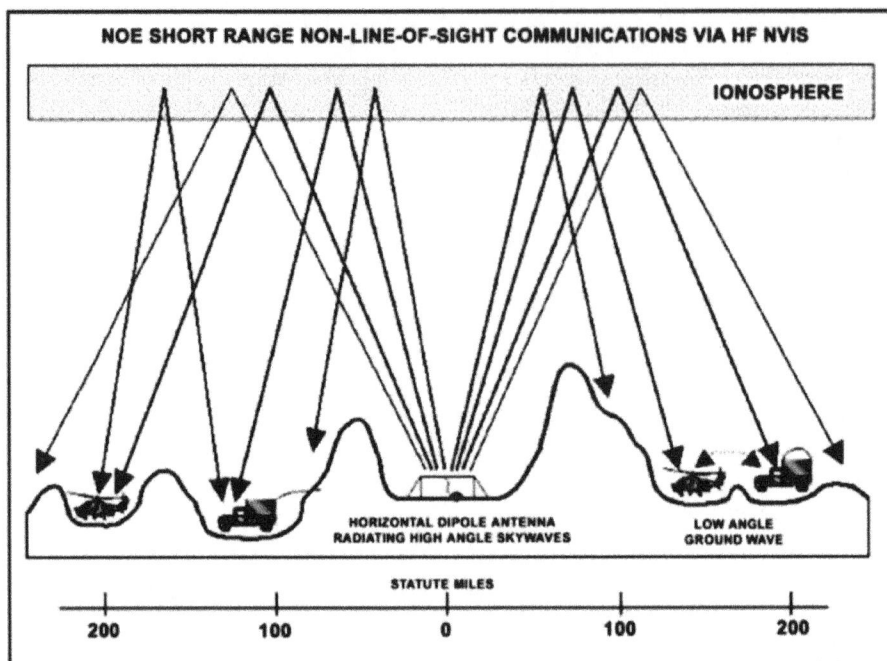

NOE SHORT RANGE NON-LINE-OF-SIGHT COMMUNICATIONS VIA HF NVIS

IONOSPHERE

HORIZONTAL DIPOLE ANTENNA RADIATING HIGH ANGLE SKYWAVES

LOW ANGLE GROUND WAVE

STATUTE MILES

200   100   0   100   200

Fig. 3
Concept of Near Vertical Incidence Sky-wave signal propagation. Horizontal wire dipoles, bent (forward or rear) monopoles (whips), and aircraft or vehicle mounted loop antennas will all produce the high angle HF radiation that was needed to solve GEN William "Scott" Wallace's problem. Operation can be fixed or OTM and since the entire signal is showered down from above via ionospheric reflection terrain is NOT a factor. Frequency selection is a critical factor since higher frequencies will penetrate the ionosphere and go off into deep space. Automatic Link Establishment features in modern radios will find the best authorized frequency to establish NVIS communications. Frequencies are normally 2-4Mhz at night and 4-8Mhz daytime for corps/theater areas of coverage.

be used since the ionosphere will not reflect signals at higher frequencies. While the U.S. Army has yet to deploy a loop antenna we do have plenty of various length whip antennas and adaptor fittings that should make OTM, BLOS tactical communications commonplace in the Army.

Myth 8 – HF doctrine does not exist in the U.S. Army.

-False again. Despite Soviet Admiral Sergei I. Gorshkov's often quoted dictum that it is "fruitless to study U.S. doctrine because they don't study it and if they did would feel no obligation to follow it". In this case doctrine is there and valid. There is a huge list of field manuals, technical manuals, military handbooks and training aids that detail solid doctrinal concepts in HF

communications that are available. On top of that there is an equally huge pile of similar doctrine in DoD and other service publications. Some of this information dates back well into the early 1920s. Lack of doctrine cannot be an excuse for S/G-6s not to employ tactical HF communications.

Over the past three-plus decades (roughly the time the Signal Center and School moved from Fort Monmouth to Fort Gordon), belief in these myths has been handed down from generation to generation of Signal Officers until "HF is no damn good" has become a mantra recited by the uninformed in order to conceal their lack on knowledge and education. This is the root cause of why Wallace was not able to emulate the performance of Guderian, Rommel and the Panzer's of the 1930s not the lack of doctrine or equipment.

Sadly, the Army has deployed across the force many of the elements that were needed to solve the communications OTM problem that Wallace so eloquently presented to both the Signal Symposium and in his testimony to Congress. More sadly still much of what was required was present and under Wallace's own command at the time he needed it so desperately. Specifically elements present that should have solved the problem were:

1 – Skimpy but adequate quantities of HF radios to provide OTM/BLOS HF communications over corps/theater size areas from corps to battalion level if anyone cared to locate them. This equipment is on the unit Table of Organization and Equipment.

2 - Doctrine that laid out the correct net radio structure to provide multiple HF communications paths from where the bridge information existed to where it was needed if anyone cared to read it.

3 – Procedures required to get the correct antennas configured and the radio equipment on the air in the OTM/BLOS mode - if anyone cared to implement them.

What was not present and caused the critical breakdown in communications was:

1 – Education- The Army no longer conducts an military occupational specialty producing course called Radio and Microwave System Officer (O505) as it did in the 60s and 70s. This hurts in an Army that uses combat net radios for almost everything on the battlefield. The idea of calling CNR "user-owned and operated" is bankrupt. Radio communications is not a trivial subject and to learn it to a level required for effective use in combat Signal officers, warrant officers and senior NCOs need to be better educated - particularly in the basic fields of radio physics/systems engineering, antennas, radio-wave

propagation and frequency engineering. These subjects cannot be given the rush treatment during initial training as we do today. Each requires hard time in the classroom. This doesn't mean everyone in Signal needs to have a bachelor of science in electrical engineering but it does mean far more instruction than we give now and at a far higher level - particularly for OTM and BLOS communications such as HF and SATCOM. Signal personnel educated this way would have in this situation been able to analyze the tactical situation and had the right CNR (HF) and the right antenna, and the right frequency assignment ready to go as the situation developed. If it is beyond the Army's current capability to educate to this level then we should consider contracting a local community college or technical school to deliver the proper instruction. This needs to be backed up with a takeaway package of technical publications and computer-based training that is retained by each graduate for use in the field. This is not new and is currently implemented by other services.

2 – Training – For far too long the Signal Corps idea of training has been teaching students what button to push, or what module/box to change. What we have failed to teach is the "why". In many cases Signal personnel cannot explain why they are doing what they are doing - they are just doing it by rote when it comes to CNR systems. A logical thinking process and a reasonable knowledge of how things work is just not being imparted to many (not all) of our signal soldiers. When these personnel get to field (S-6) assignments they are expected to be the commanders technical experts with all communications/automation equipment whether it is owned by the user or not. Often they fail with CNRs through no fault of their own because they have not seen this "user" radio equipment before. When as so often happens in deployed situations something unusual happens, the button - pushers and

the module changers are stymied because they have no training in a logical method that will isolate CNR system problems and fix them based on knowing the equipment and how it works. Additionally, personnel trained this way are not prepared to jump into situations like the one that faced Wallace with innovative technical applications to fit unique tactical situations on the fly. For many years, I have heard numerous senior Signal officers say essentially "you really never learn this stuff until you get to a unit and you're on the job". The incident at the Iraqi bridge proves that our branch "OJT" concept is as bankrupt as the "user-owned-and-operated" concept. Hoping for the best is just not a course of action that works. Hard time in the classroom backed up by field training is the only thing that does prepare a Signal soldier for combat operations.

The situation that Wallace talks about is not new. In fact it is roughly equivalent to the famous Operation Market Garden of World War-II depicted in the film *A Bridge Too Far*. In both operations higher headquarters knew the tactical situation confronting the forward force well because of air reconnaissance and similar high command resources. What failed in both cases was the supporting signal organizations ability to use on-hand, existing, CNR systems to establish radio communications between forward and supporting forces that were BLOS but really not that far away. In 1944 GEN Omar Bradley stated to his subordinate commanders after the Market Garden force had been extricated "It took an act of Congress to make you officers and gentleman it takes communications to make you a commander." Truer words were never spoken.

We failed Wallace in this instance. He knows it. It was a small but significant action in a big operation but he is focused on it. The force and the Signal staff recovered from the shortfall in combat communications and moved on to take the bridge eventually but with some difficulty. The force and the Signal staff continued the fight until we won the war. The failure was however indicative (at least to Wallace) that something was very wrong with combat communications and the Signal Corps. Wallace would not have brought this up before the Congress of the United States and again at the 2003 Signal Symposium if he were not highly concerned.

What we need to do is listen to what the general is saying and fix it with the resources we have on hand today. This includes:

1) getting more HF hardware and putting it where it needs to be,

2) building a working systems architecture for all Army organizations,

3) building unit TOEs that track the systems architecture,

4) having operations and organizational concepts that track the SA and the TOE, and

5) by dispelling the myths about HF radio systems shown above through a well thought out professional CNR education and training program,

6) by replacing in the Signal Corps the bankrupt concepts of "user owned and operated" and "on the job professional signal training" with level appropriate knowledge and experience and by providing CNR sustainment training on a regular basis to signal staffs in their field locations.

We need to remember in this age when the "technology junkies" seem to rule our thinking with their exotic networking and information transfer ideas that the simple CNR is as basic to the Signal Corps as the rifle is to the infantry. Often in Blitzkrieg operations like OIF the network centric way of fighting is out the window (Wallace talks about this in his presentation also) and the simple HF-CNR is the only system that can get the minimum essential traffic through -even if the force is moving and spread BLOS. The HF radio has the characteristics that can hold highly mobile operations together over any kind of battlefield until more elaborate higher volume

> *What we need to do is listen to what the general is saying and fix it with the resources we have on hand today.*

systems can be deployed. The famous Signal dictum of PACE (primary, alternate, contingency, emergency) is a very valid concept that includes HF and all CNRs (VHF/UHF etc.).

In order to be responsive to changing battlefield situations and maintain our credibility with the commanders, signal personnel need to understand all systems (including the humble single channel HF radio) and be able to fit the tool to the job. On the battlefield you never know what system will have to carry the ball for a commander who needs to communicate.

Mr. Fiedler – a retired Signal Corps lieutenant colonel – is an engineer and project director at the project manager for tactical-radio communications systems, Fort Monmouth. Past assignments include service with Army avionics, electronic warfare, combat-surveillance and target-acquisition laboratories, Army Communications Systems Agency, PM for mobile-subscriber equipment, PM-SINCGARS and PM for All-Source Analysis System. He's also served as assistant PM, field-office chief and director of integration for the Joint Tactical Fusion Program, a field-operating agency of the deputy chief of staff for operations. Fiedler has served in Army, Army Reserve and Army National Guard Signal, infantry and armor units and as a DA civilian engineer since 1971. He holds degrees in both physics and engineering and a master's degree in industrial management. He is the author of many articles in the fields of combat communications and electronic warfare.

ACRONYM QUICKSCAN

ALE – Automatic Link Establishment
AMEDD – Army Medical Department
AVN – Army Aviation
BCOTM – Battle-Command-on-the-move
BLOS – beyond-line-of-sight
C2 – command and control
C3 – command, control and communications
CNR – combat net radios
CP – command post
DoD – Department of Defense
DSP – digital signal processing
HF – high frequency
LOS – line-of sight
MELP – Mixed Excitation Linear Predictive
MOS – military occupational specialty
NVIS – Near Vertical Incidence Skywave
O&O – operations and organizational
OJT – on-the-job-training
OTM – on-the-move
PACE – primary, alternate, contingency, emergency
SA – systems architecture
SATCOM – satellite communications
SIDs – sudden Ionospheric disturbances
SOF – Special Operations Forces Forces
TOC – Tactical Operations Centers
TOE – Table of Organization and Equipment
UAVs – unmanned aerial vehicles

www.ingramcontent.com/pod-product-compliance
Lightning Source LLC
Chambersburg PA
CBHW081816200326
41597CB00023B/4268